THE END OF GENDER

Debunking the Myths about Sex and Identity in Our Society

DR. DEBRA SOH

THRESHOLD EDITIONS
New York London Toronto Sydney New Delhi

Threshold Editions
An Imprint of Simon & Schuster, Inc.
1230 Avenue of the Americas
New York, NY 10020

First Threshold Editions hardcover edition August 2020

Interior design by Davina Mock-Maniscalco

Manufactured in the United States of America

10 9 8 7 6 5 4 3 2 1

Library of Congress Cataloging-in-Publication Data

Names: Soh, Debra, author.
Title: The end of gender : debunking the myths about sex and identity in our society / Dr. Debra Soh.
Description: First Threshold Editions hardcover edition. | New York : Threshold Editions, 2020. | Includes bibliographical references. |
Summary: "International sex researcher, neuroscientist, and frequent contributor to *The Globe and Mail* (Toronto) Debra Soh debunks popular gender myths in this fascinating, research-based, scientific examination of the many facets of gender identity"—Provided by publisher.
Identifiers: LCCN 2020008551 (print) | LCCN 2020008552 (ebook) | ISBN 9781982132514 (hardcover) | ISBN 9781982132521 (trade paperback) | ISBN 9781982132538 (ebook)
Subjects: LCSH: Gender identity.
Classification: LCC HQ18.55.S64 2020 (print) | LCC HQ18.55 (ebook) | DDC 305.3—dc23
LC record available at https://lccn.loc.gov/2020008551
LC ebook record available at https://lccn.loc.gov/2020008552

ISBN 978-1-9821-3251-4
ISBN 978-1-9821-3253-8 (ebook)

For everyone who blocked me on Twitter

CONTENTS

THE FUTURE

THE END OF GENDER

INTRODUCTION

THE BATTLE AGAINST BIOLOGY

Hello!

Are you afraid someone saw you pick up this book? You shouldn't be.

Everything in here is backed by science and facts. None of this should be controversial. And yet it is.

On some level, you already know this. There is a price for telling the truth about gender, especially in today's day and age. I've paid it; maybe you have, too. Or maybe you are in the early stages of contemplating doing the same. Once you've opened your mouth and the words have come out, you can never go back. But you won't want to.

At first, I didn't think it was going to be this bad. I was so sure the current political climate, including the accompanying science denial, policy changes, mob mentality, shaming, and allegations of hate speech, would only be a temporary thing. The pendulum could only swing so far in one direction for so long. Certainly it would be swinging back.

And so, I waited. I'm still waiting. Instead, this false and unscientific way of thinking continues to spread beyond the silo of academia and education, permeating the mainstream press, medicine, scientific organizations, entertainment, social media, law, and tech. There are no signs of turning back. It's not a question of whether you will encounter this ideology in your life, but to what extent and how willing you are to recognize that it's happening.

Before I dive into the nitty-gritty details about the science of gender and how it applies to our lives in the context of so many issues today, I'll tell you about the thought process that led me here, from my time as an academic sex researcher to the moment I began putting these words on a page.

Few people will tell you that doing a PhD is a good idea, especially if they have one themselves. Graduate school consists of countless long nights and weekends spent working, goalposts constantly shifting, and endless bureaucracy, only to contend with an unpredictable job market upon graduating. Most graduates who leave the academy will get a job in industry, one that is relevant to their specialization. Few will start over in a completely unrelated field, like journalism, as I did.

My eventual decision to abandon the ivory tower was the result of the political climate shifting in ways I could have never predicted. In an ideal world, scientists wouldn't have to worry about the political implications of their work. Sex research is controversial by nature, and sex researchers are used to contending with interference coming from *both* sides of the political aisle. But while there is much backlash whenever the political right tries to meddle with sex science, and science more broadly, when the political left starts suppressing science, most people look the other way.

It's important to note the difference between sex research and sexology—my former field—and other academic disciplines, like gender or women's studies, because they are not the same. Sex research refers to scientific disciplines that use quantitative (numbers-based) methods, including statistics, to understand human sexuality and gender. Related disciplines include biology, psychology, neuroscience, and medicine.

When done properly, sex research is rigorous, because science is designed to eliminate bias and confounding variables, so that you know that what you found is legit. Gender studies tends to use qualitative methods, like interviews and autoethnographies (which are like diary entries). It is not a scientific discipline. Although there are definitely gender scholars who are careful and rigorous with their work, many others are not fans of the scientific method.

This battle between scientific and antiscientific enterprises plays out in meaningful ways, as we'll see. And as left-leaning science denial continues to gain a greater foothold in the academy, research challenging progressive narratives has become increasingly precarious territory. It isn't as though scientists are intentionally publishing controversial research findings with the desire to upset and offend people. But the fear of potentially discovering something that hasn't been given the progressive stamp of approval has certainly become a larger factor influencing the types of questions a researcher chooses to pursue or avoid. Contrary to what you might expect, most sexologists (including me) who oppose these winds of political change are liberals.

Don't get me wrong—I loved my research and sexology had become my intellectual home. At the time, I was using brain-imaging techniques, including functional magnetic resonance imaging (or fMRI), to better understand paraphilias,* sexual orientation, and hypersexuality in men. As someone who is surprisingly old-fashioned and vanilla (that is, nonkinky) in my personal life, one of my biggest aims as a sex researcher was to combat sexual stigma and shame.

The laboratory I worked in was the only one in the world doing this kind of research and I felt like I had won the lottery. But things had changed so much during my time in graduate school that I found myself at an impasse. I decided to leave behind my dreams of one day starting my own research lab to instead set my sights on an entirely new career as a journalist so that I could be free to speak my mind.

---

*paraphilias: unusual sexual preferences.

Transgender Children

In my decision to leave academia, writing an op-ed became the catalyst. I had noticed a trend in mainstream news stories that left me feeling unsettled because they were extremely one-sided. An endless stream of pieces presented glowing stories about children as young as age three transitioning—changing their haircut, taking on a new name, and championing the use of medical interventions to halt some physical changes and facilitate others.

From as young an age as their parents could recall, something was *different* about these kids. They would say things that would make any parent's heart wrench, like they were born in the wrong body, that "God made a mistake," and that they wanted to die and be brought back to life as the opposite sex. These children would be suffering immensely until they were allowed a gender transition.

Not only that, but when I scrolled down below these articles, I would see parents posting in the comments section. Many would say that they, too, had a child who felt they were born in the wrong body, but the parents were unsure as to whether this direction was right for them. They were uncomfortable with the idea of allowing a child to essentially be a guinea pig, undergoing such a new and experimental approach. These parents would clearly state that they were in support of the transgender community, but were torn as to what to do.

In response, other commenters would attack the parent, calling them transphobic and bigoted, saying that their child would kill themselves and they would be the one to blame. I could only imagine how a parent would feel, voicing genuine concern for their child's well-being and in turn being harassed and dehumanized for merely asking a question. I could also see how this intimidation could lead to more parents allowing their children to transition.

After seeing this happen again and again, one story after another, I felt I had to speak up, for the sake of these parents and their children, because so few of my colleagues were willing to say anything.

This silence was the result of a long, ugly history between transgender activists and sexologists, in which activists would go after sex researchers if they didn't like a particular study that had been done or what an expert had publicly said. Everyone in the field who criticized transgender ideology would be attacked ruthlessly, to the point, in some cases, of nearly having their professional and personal reputation ruined (see Chapter 4).

I wrote an opinion piece that countered the early transitioning narrative, then deliberated about publishing it for about six months. Even though the science I referenced was solid, it was too politically volatile, and drawing attention to it would be the equivalent of expelling myself from academia.

I had asked my colleagues if, in their opinion, I should wait until I was a tenured academic to have it published. Once I had tenure, I would have job security and institutional support backing me, so the pushback from those who were enraged would, in my mind, be buffered.

One of my mentors, whom I trusted dearly, having known him since my first year as a graduate student, offered a wake-up call.

"Tenure won't protect you," he said.

It solidified my decision.

I came out of a research meeting one afternoon to see that the piece had been published on *Pacific Standard*'s website. And then, as if on cue, the mobbing began.

I do not, by any means, live my life online. I like being in the real world, talking to people in real life. A good day, in my books, is one in which I don't have to touch my phone. When the mobbing for this particular piece started, I hadn't yet acclimatized to the culture of these platforms or how to properly game a response. All I knew is that I had taken aim at a sacred cow and people were very, very angry.

A wise person once said you should never read the comments, but I always do, not because I'm a masochist, but for the rare occasion that I learn something. Usually, responses are off topic, ranging from predictably confrontational (like the person who called me a "Nazi c*nt") to whimsically creative ("She looks like a ladyboy in a bad wig"). Still, occasionally I will uncover something of value that changes the way I think or gives me perspective. Over the years, I've also done a good deal of digging into arguments against mine, because I figure it's not possible to know whether you're right without understanding the other side.

In the Internet age, activists and allies are not content at seeking justice. They seek to punish heretical thought. They want your head on a stick, for you to die in a fire, or preferably, both. If you've never been mobbed on social media, this is what it feels like: One or two notifications will trickle in at a hostility level of 10 out of 10. For me, they consisted of parents accusing me of being pro–conversion therapy and wanting their trans kids to kill themselves.

For a minute, you'll think, *People seem pretty mad. But maybe they'll get over it.*

Then, you'll be flooded. You won't be able to refresh your notifications fast enough. For some reason, every time I refreshed the page, I foolishly thought, *Surely someone will say something nice in between people telling me I should be hanged.* A few brave souls did try calling for a gentler approach. They were immediately pounced on and drowned out by the mob. Some of my defenders went so far as to offer contrition, thanking their mobbers for "educating" them and apologizing for defending a transphobe.

Once you've survived your first mobbing, every subsequent one is a breeze. In my case, knowing that the backlash was coming was a blessing. I can't imagine what it would feel like to cross trans activists without realizing what you had done. A friend of mine was similarly mobbed several years later for stating on social media that only women could get pregnant. After sustaining several days of online persecution, he told me, slightly shaken, "I'm never tweeting about trans issues again."

It may sound like I was alone amid the mess of that mobbing, but I wasn't. Much to my shock, conservative media had come to my defense. Much like radical feminists and fundamentalist Christians joining hands over the transgender bathroom debate—another alliance I could have never anticipated (see Chapter 6)—the unforeseen shift in the zeitgeist has led me, a sex-positive, nontraditionalist, gay rights supporter who once made a living studying kinky sex and sex toys, to be considered, by those who negated what I had to say, as right-leaning and Republican. I found it quite funny when David French, whom I respect a lot, wrote positively about my opinion piece for *National Review,* noting that I was "certainly no cultural conservative," while linking to an article I had written about furries (a subculture of people who identify as anthropomorphic animals).[1]

A year later, I began writing my first regular column as one of Playboy.com's resident sex scientists. A year after that, I graduated with my PhD in the neuroscience of sex. With everything I've witnessed since I left, I no longer question whether I made the right decision.

Some people will be mad at me for writing this book. I don't care. I wrote it to answer your questions at a time when it's next to impossible to tell apart politically motivated ideas from scientific truth. Gender has transformed into a cultlike concept, and public knowledge has been overturned to reflect pleasantries that affirm the feelings and beliefs of particular groups. Scientific research is no longer about exploring new ground, but promoting ideas that make people happy.

So many of you have asked me how to go about countering misinformation when you are debating colleagues, friends, and family members. What does the research say about a particular issue? How can you tell whether a scientific source is accurate and unbiased?

Each chapter debunks a particular myth in the contemporary gender debate, going against much of what we are currently being fed from academic experts, educators, scientific publications, and the media; for example, I challenge the erroneous belief that sex and gender are socially constructed; that there is an infinite number of genders; that young children with gender dysphoria should be allowed to transition to the opposite sex; and that trans women are quintessentially no different from women who were born women.

Other taboo truths include how transitioning is influenced by sexual orientation and, for some people, sexual arousal; how biological sex differences lead to differences in dating and sex; and that biology has a greater influence on gender than child-rearing. I also discuss why academic freedom is so important, particularly within sexology and the biological sciences, at a time when academics are being treated as pariahs for simply doing their jobs.

For each topic, I have presented a comprehensive summary of the available research literature, so you can draw your own conclusions and make up your own mind. My hope is that you will be inspired to become defenders of science in your own lives.

Leaving academia after eleven years turned out to be one of the best decisions I ever made. It has afforded me a mental tranquility that I could not have had, had I stayed. I no longer have to choose between telling a lie or staying quiet. It is freedom that other scientists and academics are not afforded in the current climate.

As a journalist, sometimes it feels as though you're screaming into a void. You forget that people actually read your writing. To every one of you who has shared your stories with me, whether it has been online or at speaking events or when you bump into me at the airport or on the street, I have taken all of it into account and am grateful for it. I know what it's like to feel alone, to be looking around the room in disbelief at everyone else who is nodding along at what's being said and asking yourself, "Am I the only one who feels like this?"

You aren't, and I'm here to tell you why.

THE FOUNDATION

MYTH #1

BIOLOGICAL SEX IS A SPECTRUM

What does it mean to *feel* like a woman? Or to feel like a *man*?

Who could have predicted that such benign questions would evolve, seemingly overnight, into the next civil rights frontier? Once uncontroversial and matter-of-fact, discussions about gender have become a cultural minefield, hostile terrain striking fear into the heart of anyone who treads against orthodoxy.

In an era in which computer keyboards have replaced physical pitchforks, a zero-sum game has unfolded in real time, in which playing along wins you favors and accolades and moves you ahead in life, and questioning the narrative means you will be socially shunned at best, and harassed, threatened, fired from your job, and fearing for the safety of your family, at worst.

The consolation prize, however, will take the form of knowing you were right all along. A price will be paid for the glamorization of false information, including the gaping sense of alienation between the political left and right; the subsequent backlash when women realize that so many cultural messages targeting them have been lies; and the emotional fallout and class-action lawsuits resulting from public policies that sacrificed the well-being of children and vulnerable people in the name of a larger activist goal.

I hear from individuals from all walks of life who want to learn more about gender so they can form opinions in a factual, unbiased way about the issues that affect their lives. They don't have nefarious intentions, and because they aren't ideological, they're frustrated by how impossible it's become to find straightforward answers backed by science.

Adding to their dismay, in many cases, is realizing that much of the information that claims to be scientific is partisan and factually inaccurate. They also can't understand how on earth we've gotten to a place where simply seeking this information is interpreted as threatening and leads to accusations that a person is sexist, misogynistic, transphobic, and hateful.

Sex Versus Gender

A good place to start is defining basic terms, like sex and gender. These terms are both related to one another and distinct, and because of this, much confusion often ensues.

Biological sex is either male or female. Contrary to what is commonly believed, sex is defined not by chromosomes or our genitals or hormonal profiles, but by gametes, which are mature reproductive cells. There are only two types of gametes: small ones called sperm that are produced by males, and large ones called eggs that are produced by females. There are no intermediate types of gametes between egg and sperm cells. Sex is therefore binary. It is not a spectrum.

By contrast, *gender identity* is how we feel in relation to our sex, regarding whether we feel masculine or feminine. *Gender expression* is the external manifestation of our gender identity, or how we express our gender through our appearance, like clothing and hairstyle choices and mannerisms.

Similar to sex, gender—both with regard to identity and expression—is biological. It is *not* a social construct,* nor is it divorced from anatomy or sexual orientation. Despite what contemporary scholars may have you believe, all of these things are very much linked. Biology, not society, dictates whether we are gender-typical or atypical, the extent to which we identify as the sex we were born as, and the partners we are sexually attracted to.

*social construct: the product of one's social environment or learning.

When sperm fertilizes an egg at conception, the baby will be either female or male. This biology will influence hormonal exposure in the womb, as well as the child's resulting gender identity. At about seven weeks, if the embryo is male, the testes will begin to secrete testosterone, masculinizing the brain. If the embryo is female, this process does not occur.

There are thousands of studies showing the effects of prenatal testosterone on the developing brain. In fact, this exposure to testosterone has a powerful effect on the ways in which male and female brains grow. In a 2016 study in *Nature's Scientific Reports*, researchers at the University of California, Los Angeles, found that testosterone exposure alters the programming of neural stem cells responsible for brain growth, leading to differences between the sexes before the brain has finished developing in utero.[1]

From a scientific perspective, gender identity is basically synonymous with biological sex. There are, of course, exceptions to the rule, including intersex people and people who are transgender (whom we will discuss in a moment).

In a time not too long ago, biological explanations were used to suggest that women were bad at math and belonged in the kitchen. A woman's only value was bearing children and helping her husband succeed. Thankfully, times have changed, but biology remains stigmatized, unable to shake its former reputation. It is seen as inherently sexist, and even contemporary studies using biologically based methods are being dismissed as antiquated without a second thought.

In an attempt to correct for past wrongdoing, scientific organizations, new academic studies, and experts in the field have taken to overcompensating, actively burying any links to biology's relevance and claiming the scientific consensus says one thing when it actually supports the opposite. Biology has been equated with bigotry and is, in turn, being scrubbed from existence. In popular culture terms, biology has been *canceled*.

Watching the pace of this science denialism spread—including claims that there is no such thing as being biologically male or female, or that "biological sex" isn't a coherent concept[2]—has been astounding. In today's climate, gender has been branded as an ephemeral, intangible thing, something that can't quite be described or explained beyond one's personal experience and self-identification. Biological sex is now, worrisomely, following suit.

These efforts, however, are unnecessary and shortsighted, because denying biology will not help us live more productive, meaningful lives. Instead, hiding biological facts only sends us back to the Dark Ages, to stumble around, rediscovering what we already know. As we shall see in the coming chapters, the issue should not be with what science tells us, but how these findings are used.

Although sex and gender are both biologically based, it isn't accurate to use them interchangeably. Nowadays, *gender* is used almost exclusively, even when a person is actually referring to *sex*. When a baby gorilla was born at a public zoo in Toronto, journalists happily shared the news that the gorilla's *gender* would be announced shortly. But animals, including intelligent ones like gorillas, don't have a gender. They have a sex.

As an anecdote to illustrate why it's important to make this distinction, I'm reminded of a time I was invited to host a workshop at a large research institute. Whenever I'm invited to nonpolitical events, even though the organizers have explicitly reached out to me, I have a tendency to assume that everyone in the audience, as cordial as they may be, upon reading my work secretly thinks I'm the devil.

Upon arrival, one of the research scientists I met pulled me aside to introduce herself and to tell me about one of her experiences. To my surprise, she spoke of her frustration regarding how gender ideology was influencing the kinds of findings she could report in her research.

Reputable scientific journals require any study submission to be peer-reviewed by other academic experts before it can be published. One of her manuscripts, she said, used an animal model. Mice. As part of standard protocol, she described the mice based on whether they were male or female. The journal responded that in order to publish her paper, she would have to replace every instance referring to an animal's "sex" with the word "gender."

"I don't know what the mouse's *gender* is," she sighed. "It hasn't told me."

She couldn't understand why the journal thought this change was necessary, as use of the word "gender" would be, from a scientific perspective, incorrect. It seemed like an example of falling into ideological lockstep—"sex" was verboten because it is biological. It didn't matter if it was the correct word to use. "Gender" had become the accepted nomenclature. These cultural politics mattered apparently, even when referring to rodents.

Another common example is when excited parents throw so-called gender-reveal parties. They're not really revealing the baby's *gender*, but rather, the baby's *sex*. I can understand why, whether or not a person knows the difference between the two, they might prefer to *not* call it a *sex*-reveal party. Something about using the word "sex" in reference to a child feels a bit disconcerting.

As a former sex researcher who has witnessed the full range of responses that people have upon learning about my former vocation (ranging from "Oh wow, that's so fascinating!" to losing color in their face and disappearing without saying another word, due to embarrassment or discomfort), use of the word "gender" over "sex" has likely also been embraced because "sex" connotes sexual intercourse, and human sexuality remains stigmatized. At the same time, "sex" is a clinical word, and like any word, it shouldn't be taboo or frowned upon to use it.

Sexual orientation influences gender (see Chapter 4), but gender is not synonymous with one's sexuality. Research has shown that people who are gender-atypical (who look and behave more like the opposite sex than people who share their sex) are more likely to be gay. Being transgender, however, is not an indication of whether someone is gay or straight. And being gender-atypical is not an indication that someone is necessarily trans.

By definition, a transgender person feels that their gender identity is more in alignment with the opposite sex than their birth sex (which refers to their sex at birth). The word "transgender" encompasses the Latin prefix *trans-*, which means "on the other side of." (The word "cisgender" is sometimes used to refer to people who are not transgender and who identify as their birth sex. It similarly uses a Latin prefix, *cis-*, which translates to "on this side of," or roughly, one's gender identity and birth sex are on the same side.)

Transgender women are often assumed to be gay. This is likely because wanting to become a woman is understandably seen as feminine, and feminine men tend to be sexually attracted to men. In sexology (the scientific study of sex and gender), when someone transitions to the opposite sex, their sexual orientation refers to whom they are attracted to when taking into consideration their birth sex. For example, a transgender woman (someone who was born male but identifies as female) who is attracted to men is considered, in sexological terms, to be gay, since her birth sex is male and is the same as the partners she's interested in. If she were attracted to women, she would be considered straight. But by virtue of being transgender, this says nothing about whether a person will be attracted to women, men, or both sexes. By the same token, being heterosexual does not mean that someone is not trans.

Biological Sex Is Not a Spectrum

For more than 99 percent of us, our gender *is* our biological sex. Regarding the 1 percent of individuals for whom gender identity and biological sex don't align, they may identify as transgender or possess a medical condition known as intersex. The latest statistics show that 6 in 1,000 American adults identify as transgender and as many as 1 in 100 people are intersex.[3] In some cases, an individual may be intersex and also identity as trans, but it's important to note that not all intersex people are trans and not all trans people are intersex.

Intersex is also known as having a difference of sex development, and was previously known as hermaphrodism (which is now considered a stigmatizing and insensitive term). Someone who is intersex possesses reproductive or sexual anatomy that would be considered atypical because it doesn't fit the standard definition of male or female. An example would be someone who has both a vulva and testicular tissue. Intersex conditions occur as a result of differences in genetics and hormonal levels in utero. As many as thirty variations exist.[4]

Regarding the role of chromosomes in human development, all eggs have an X chromosome, and each sperm carries either an X or a Y. Upon fertilizing an egg, the sperm's X or Y chromosome combines with the X chromosome of the egg. Women usually have XX chromosomes and men usually have XY chromosomes.

Some people with intersex conditions will have different sex chromosomes, like in the case of Klinefelter syndrome, wherein a man will inherit an extra X chromosome from either his mother or father. In other cases, someone who is intersex may have the chromosomes typical of one sex but outwardly appear as the opposite sex. For example, girls with an intersex condition called androgen insensitivity syndrome have XY chromosomes and male internal organs, but their body can't respond to testosterone, so they will appear female. And some intersex people have chromosomes that are typical of their sex. For instance, girls with congenital adrenal hyperplasia, another intersex condition, experience masculinization in the womb and have XX chromosomes typical of females.

Those advocating that biological sex is a spectrum frequently tokenize the intersex community as evidence for their claims. They will say that because some individuals are born with a mix of sex characteristics, sex is not binary. This is inappropriate for several reasons. (Chapter 3 disputes the false notion that intersex people prove gender is a spectrum.)

As you'll recall, sex is determined by gametes. Intersex people tend to produce one of the two types, or are infertile. The difference lies in the fact that, in some cases, an intersex person's gametes are not in alignment with the sex they identify as. For example, girls with congenital adrenal hyperplasia are exposed to unusually high levels of testosterone in the womb. When they are born, they may have genitalia that are ambiguous, such as a clitoris that is longer than average or labia that look like a scrotum. They have ovaries that will produce eggs, typical of girls who are not intersex. But as a result of the masculinization process during development, which influences not only a person's anatomy, but their psychology, too, they may identify as male in adulthood. The fact that they have female gametes is not indicative of their sex; however, this is not to say that sex should be categorized as something wholly different if someone is intersex.

Regarding whether it would be possible for a person to produce both types of gametes, they would need to possess both ovarian and testicular tissue. Individuals with a condition known as ovotestis do possess such a combination. In most cases, however, only one type of tissue is functional; their ovaries will produce eggs, but their testes are unable to produce sperm. This condition is extremely rare, occurring in 1 in 20,000 births. Regarding the question of whether an individual is capable of producing both, one case study documented a man with ovotestis who was believed by physicians to have produced eggs at one point in time, before his testes began producing sperm.

Intersex is said to be as common as green eyes or red hair, but this is not true. An estimated 2 percent of the world's population has green eyes[5] and 1–2 percent has red hair,[6] compared with 1 percent of people in the world or fewer with an intersex condition, and an even smaller percentage who could have the potential to produce both gametes. I understand the purpose of making these comparisons; it's to say that physical features that seem uncommon are still more common than one might think, and a person may actually know or come across intersex people in their day-to-day life and not even realize it. Two percent of the population sounds rare, but surely we all know a number of redheads or people with eyes that are green. But even if intersex conditions *were* as common as either of these physical traits, they would still be considered, on the whole, as few and far between.

It thus becomes a question of whether an occurrence that is statistically rare in the general population should be considered typical, as the vast majority of us fall unambiguously into one category of sex or the other. An analogy that helps to illustrate this point is the fact that most of us have ten fingers. Some people have fewer or more than ten digits on their hands, but this hasn't led to a reconceptualization of how many fingers human beings have.

It isn't necessary to redefine "sex" or eliminate the categories of "male" and "female" in order to facilitate acceptance for people who are different. Whether we consider sex to be binary or a spectrum has no bearing on intersex individuals' right to bodily autonomy. Indeed, the intersex community has faced a long history of doctors imposing their views as to what their bodies should look like. A child should be allowed to make these decisions for themselves when they are of an appropriate age. It's possible they will be perfectly content not undergoing any surgery at all.

Doctors back then believed that so long as surgery was done early enough and everyone behaved in accordance with the decided sex, the child would grow up not knowing the difference. In Chapter 8, I go into greater detail about John Money and what happened to one of his patients, David Reimer. Money believed a child's gender was malleable and unrelated to biology. Reimer was not intersex, but his story speaks to the fact that it isn't possible to mold a child's internal sense of gender to match external surgery.

We can, and should, advocate for the rights of intersex people and anyone who does not fit neatly into male or female categories or typical gender norms. Intersex people should have the option of being legally recognized as the sex they'd prefer. They should be allowed to change the sex marked on their birth certificate, particularly if they are unaware of their condition until later in life. For instance, starting in 2019, the state of New York has allowed for individuals to have the designation of "X" instead of "male" or "female" on their birth certificate without requiring a letter from their doctor. Parents have also been granted the ability to choose an "X" for their newborns' documents. Oregon, California, Washington State, and New Jersey offer a similar designation.

This would seem like an adequately straightforward solution, but activists, many of whom are not intersex themselves, continue to push for a spectrum-based conception of sex. The most prominent example of the "sex is a spectrum" narrative reaching the mainstream occurred in October 2018, when a leaked memo from the U.S. Department of Health and Human Services sparked allegations that the Trump administration was seeking to erase transgender people.

You may remember what happened. The *New York Times* reported that the memo proposed legally defining sex as "either male or female, unchangeable, and determined by the genitals that a person is born with," under Title IX, the federal civil rights law that bans sex discrimination among government-funded institutions and educational programs. This was seen as undoing the Obama administration's previous decision to allow students access to single-sex spaces, like bathrooms and locker rooms, and sports participation based on the gender they identified as, as opposed to their birth sex. In response, a petition signed by more than 2,600 scientists, including nine Nobel laureates, proclaimed that science showed gender is not binary, determined at birth, or based on genitalia.[7]

What? The issue wasn't so much HHS's definition, but whether journalists understood the difference between sex and gender, and that it isn't appropriate to substitute one term for the other, particularly when referencing anatomy. The HHS memo was referring to sex, but media outlets, including the *Times*[8] and the *Guardian,*[9] claimed it was referring to gender. Saying that *gender* is determined by genitalia at birth conveys a much different, more controversial meaning.

As discussed earlier, *sex* is determined by gametes, as opposed to a person's genitalia. In the case of someone who identifies as transgender, sex remains male or female and is unchangeable from birth, since transitioning doesn't change a person's sex; it only changes their gender. For people who are intersex, as you now know, their gametes (or anatomy, for that matter) may or may not reflect their true sex. Use of the word "gender," however, would render the statement entirely incorrect, because gender is not necessarily determined by one's genitalia at birth. Saying so suggests that both transgender and intersex people don't exist, as they would be forced to identify as their sex at birth.

Perhaps predictably, much of the coverage referred to "gender" and not "sex," stoking a great deal of unnecessary outrage and fear. Indeed, acknowledging a transgender person's sex at birth feels insensitive, particularly for those of us who are not transgender. But the way to end discrimination against people who are gender-diverse is not to pretend that biological sex doesn't exist. The government should be able to collect basic demographic information as detailed by the memo so that it has an idea of who its constituents are, without running the risk of inciting the next civil war.

In the long run, scientific illiteracy won't help to protect these communities. If we learned anything from the HHS debacle, it's that those who claim to be pro-science, including scientists themselves, will gladly throw science under the bus when it suits their political goals.

Biological Sex Is Not a Social Construct

"Gender is a social construct" has become a ubiquitous battle cry of wokeness,* and those uttering it, not content to quit while they're ahead, have gone one inexplicable step further, claiming that biological sex is *also* socially constructed. In their words, sex is defined not by biology, but by culture and political influence, which is a curious thing to say, seeing as how this is physically impossible.

If anyone is wielding culture and politics as a way to define sex, it is these scientific revisionists. Yet their ideas continue unfurling in society, with no end in sight. This is the logic behind terminology like "sex assigned at birth," adopted by organizations like the Canadian Pediatric Society. In a recent guideline that was created to help doctors and parents understand the concept of gender, the society

*wokeness: values that are super-progressive and attuned to social justice.

told pediatricians that not only is a child's sex "assigned" at birth, but that gender identity and gender expression are unrelated, and gender is a spectrum.

The term "sex assigned at birth" stems from a desire to acknowledge the intersex community and that sometimes a person's sex is inaccurate because the doctor got it wrong. While it is a respectable goal, it needlessly gives the impression that sex is not an objective attribute and that a doctor's estimation is completely arbitrary, when in actuality, their guess will be correct, statistically speaking, ninety-nine times out of a hundred.

This kind of thinking has opened the door for the term "gender assigned at birth" to follow suit, ideologically holding hands with the false claim that gender is not biological. The underlying reasoning for this, presumably, is that if both sex and gender are "assigned" at birth, this further justifies biology's obsolescence and allowing self-identification to supersede all else.

This has spawned an entire lexicon meant to replace any reference to biological sex, including "assigned male at birth" and "assigned female at birth" and their corresponding acronyms, "AMAB" and "AFAB." Some will choose to insert the word "coercively" in front of either to account for people who are intersex and trans (as in, "coercively assigned female at birth" or "CAFAB"). Others have, however, argued that calling sex assignment "coercive" when applied to trans people does a disservice to intersex people—with trans people, doctors made an honest mistake based on the information available to them (that is, their sexual anatomy), whereas intersex people had no say in the surgical alterations they were forced to undergo to appear more female or male. There is additionally "female assigned at birth" ("FAAB"), "designated female at birth" ("DFAB"), all of the male versions of these terms, and "birth-assigned sex."

If your head's already spinning from this terminology, don't worry. You aren't the only one. But questioning whether any of it is necessary or useful, or arguing that one's sex *isn't* assigned at birth but is based on an observable measure, will have you painted as suspect and as having a preoccupation with what's between a stranger's legs.

The sex and gender binaries are considered not only pseudoscientific and wrongheaded, but also oppressive, due to the belief that they force people into stereotypical gender roles. But there is no reason why someone who was born female can't behave or present herself in a masculine way, and push back against these stereotypes while identifying as a woman. The same can be said for feminine men.

Those of us who do not fit into typical ideas of "female" and "male" do not need to be classified as an entirely different sex or gender simply because we are atypical. (Despite appearing very feminine, I am and always have been male-typical.) Also, just because someone's gender identity (female) doesn't match their gender expression (male) doesn't mean the two aren't related. The relationship between the two can be explained by hormonal exposure in the womb. I'll elaborate on all of this in Chapter 2.

Others in favor of the "sex is a spectrum" argument will say that a woman who has reached menopause, who is no longer producing eggs, is still considered a woman, and by the same token, a man who has had his testicles removed is still considered a man, so biological sex should be based in something other than gametes. This is an absurd argument, a case of activists taking issue with a concept because it doesn't serve their particular agenda. No matter how we define the concept of "woman" or "man," there will always be exceptions to the rule, which, regardless, don't justify redefining the term.

Activists wish to reinvent sex in this way because it allows for the separation of sex from an objective basis in reality. Adopting this position allows for science and logic, which currently serve as barriers to their unsupported theories, to be taken off the table when discussing gender. I would be completely open to redefining a term if it made sense to do so, and I'm sure many scientists would agree. But in this case, activists have decided what they want the outcome to be and are reverse-engineering facts to facilitate that purpose.

As we will see in Chapter 3, the fact that transgender people identify as the *opposite* sex offers further evidence that sex (as well as gender) is binary.

Miseducating the Next Generation

A few summers ago, I was invited to give a talk about the science of gender. It was my first speaking engagement as a journalist and the venue hosting myself and three other panelists would be only a few blocks from where I was living at the time. The trend of deplatforming* was already in full swing. Notable news headlines around the same time pertained to political scientist Charles Murray's appearance at Middlebury College, wherein a professor hosting him wound up concussed and in a neck brace after being violently assaulted, and conservative political commentator Ben Shapiro's talk at the University of California, Berkeley, which cost the university $600,000 for security.

All too predictably, about a month before the event was scheduled to take place, a single, anonymous person called the venue to complain. We were branded as "racists" and "Nazis," despite the fact that two of us on the panel were ethnic minorities and our event had nothing to do with race. The event's social media page was involuntarily shut down, and I don't think any of us were terribly surprised when we learned we'd been ejected from our location.

The organizer was thankfully able to find us a new venue—a beautiful, brightly lit Christian church. Now, I am not a religious person. I've been an unwavering atheist my entire life. The thought of speaking in a place of worship as an unapologetic sexpert felt like a violation of obscenity laws. But this is what happens when a group of authoritarians try to bury the truth—unexpected connections are formed so that knowledge will prevail. And I was grateful that the venue had stepped up, taking on the risk to host us, along with a fleet of menacing-looking security guards who would be keeping us safe. Our audience had

*deplatforming: shutting down an event through harassment, threats, and/or exorbitant security fees because individuals—or a single, loud individual—doesn't like what a speaker has to say.

increased to nearly one thousand people after others had heard about the failed attempt to silence us, and I wasn't about to cancel on them.

The event was a huge success. It was also one of the first times I had gotten the chance to meet my audience and hear their thoughts. One of the highlights of the evening was when a pastor shook my hand, saying, "I never thought I'd be thanking a sex columnist, but here I am." For so many of those I met, it didn't matter which side of the political spectrum they were on or what their personal beliefs were. They just wanted to hear what the science actually said.

I have felt that same sense of relief in my own life, whenever I'm reading about a contentious issue in hopes of better understanding it, and wanting desperately to find sources that are impartial, when every single one I uncover seems biased, either to the political left or to the right. I understood how they felt. Science shouldn't be like that; it should always be bipartisan.

One of the themes that stood out to me that night, after talking to many parents in the audience, is how strongly gender theory has taken hold of the education system. Children are being recruited as pawns to promote a manner of thinking that serves only adults. Parents can no longer trust that their child's school will teach them a curriculum based in facts. Over the years, I've received countless screen shots of guidelines and class handouts from concerned parents, caregivers, and teachers themselves. I question whether other parents realize this is happening, or worse, that they do, but don't care.

In the province of Ontario, Canada (where I am based), the sex education curriculum in schools is comprehensive, as it should be, teaching children about contraception, anatomy, sexually transmitted infections, and consent. An abstinence-only curriculum, on the other hand, teaches children to avoid having sex until marriage.

Contrary to what one might expect, comprehensive, science-based sex education actually improves young people's decision-making about their sexual health. They are more likely to delay becoming sexually active and to use condoms and other forms of contraception when they do. Comprehensive sex ed programs lower the risk of teen pregnancy compared to abstinence-only programs or a lack of sex education altogether.[10]

All of that would be fine and well except that the curriculum veers off into questionable territory. As part of the curriculum, kids are taught about puberty, but instead of referring to children as simply "boys" and "girls," some teachers refer to them as "people with a penis" and "people with a vagina." Other parents I've spoken to have told me about how their children's general practitioners will refer to "girls that have a penis" and "boys who have periods." This is presumably in reference to intersex and transgender people, and an attempt on the part of educators and doctors to correct for past mistakes. As thoughtless as it was to once claim that trans people don't exist or that performing unwanted surgeries on intersex kids is acceptable, this vocabulary stems from an ideological vision that will only confuse and harm children, as we will see in the case of rapid-onset gender dysphoria.

Other key points include telling children that people can identify as both genders or neither, that gender is due to traditional stereotypes, and that gender identity and sexual orientation are completely unrelated. Updated sex ed guidelines in the state of California have followed in a similar vein, teaching children as early as kindergarten about gender fluidity.[11]

This is a theme we will see over and over again—targeting children with these ideas at a young age, when they don't know any better. The indoctrination continues even after they've graduated from high school. For those going on to higher education, many campus health websites have entire sections dedicated to gender, sex, and sexuality that are replete with incorrect information. For example, it isn't uncommon to read that biological sex can change over one's lifetime, that gender is a social construct, and that sex, gender, and sexual orientation are spectra. This only scrapes the surface of what is formally taught for course credit in university.

I worry about what biology textbooks will look like in a few years, when older editions are updated to reflect this way of thinking. Biology has been described as a "scientific construction" by some left-leaning media outlets, which has left me wondering what exactly this means. I am guessing, since science can never be definitively proven—only tested and retested in order to determine whether a particular idea continues to stand—this means scientific concepts and findings thereby have no merit. Yet those critical of the scientific method have zero alternatives that could adequately take its place.

Part of the problem is that many of those denying biology lack basic science literacy. My sense is that individuals supporting purely social constructionist explanations do so because they've had a bad experience with science. Perhaps they find it uninteresting or they can't understand it, so they turn to other schools of thought.

It also doesn't help that most research publications are concealed from the public behind a pay wall. Even if they are accessible, they are filled with jargon that is difficult to comprehend. When science is seen as intimidating, it will be avoided. If it is simultaneously hidden, this compounds the problem because it becomes easier to ignore.

You cannot know whether your perspective is correct without considering arguments against it. It's become commonplace nowadays for progressives to refuse to engage with those with whom they disagree because it's seen as legitimizing the enemy or a waste of time because they've already decided who's right. Pejoratives like "biological essentialism" and "scientism" continue to be thrown around, paired with unwavering conviction in proclamations like "I don't need to read that study because I already know what it's going to say."

As a result, debates about controversial scientific findings backing biological explanations almost always consist of scientists who know what the research literature says, and activists who haven't read a single study, denying them. (In other cases, so-called scientific experts will take up the activist cause, as we'll see in Chapters 2 and 9.)

I don't believe biology should be seen as threatening, but I can understand why some think it is. To some, biological explanations transform a colorful existence into one that is unremarkably black-and-white. They conjure up the feeling that we are somehow not in control of our lives or the decisions we make, but this is untrue. We can still lead meaningful lives while acknowledging that some aspects of who we are are predetermined.

My training as a scientist taught me many valuable lessons, but the one that has stood out to me the most, the deeper I wade into the culture war, is that there are two ways to live in the world: seeing it as it is, or seeing it as you want it to be. It's one thing to be ignorant about an issue, and another to be confident in one's ignorance. It's the difference between taking a piece of information at face value, and taking the time to look up the primary source to see whether it's true.

Our opinions don't define us, but they can feel like an extension of who we are, and being challenged about them can be upsetting and disorienting. We've all been there. Sometimes disagreement can come across as a personal attack, even when the other person has the best of intentions. Engaging in a good-faith conversation requires effort and coming to terms with the possibility that we may be wrong.

Some prefer to live in ignorance because it's comforting. And some of us prefer to live in reality, despite its greater cost. The truth can be suppressed, but it will always come out. In this case, when it does, you will be ready.

MYTH #2

GENDER IS A SOCIAL CONSTRUCT

When I was in graduate school, I was very, very feminist. I would get into verbal altercations with strangers on a regular basis—at school, at work, on public transit, you name it—about women's rights and the subtle ways in which society encouraged our subjugation. I believed gender was a social construct and that biology was inherently oppressive. I would argue with the men in my life about this. Never in a million years did I think I'd one day be advocating publicly for the importance of sex differences. Back then, I would have accused women who say the things I say today of having internalized misogyny. It was only when I began studying sexology that I realized that gender is biologically based, and that the denial of sex differences isn't necessary for gender equality.

Mainstream feminism has been very effective in spreading its tenets about gender and is now sowing the fruits of its labor. In the name of affording women and girls the same opportunities and rights as boys and men, news stories, educational institutions, and governmental policies have taken to broadcasting a similar message: men and women are, at the core, the same, and any differences we do see are due to socialization and sexism.

Ask any young feminist whether gender is biological or socially constructed and she will proudly and emphatically tell you it is the latter. Men are terrified to say anything that could even be construed as challenging this, or any aspect of feminist doctrine, out of fear of being labeled a Neanderthal—or worse, a misogynist.

The myth that gender is a social construct first gained prominence through writing from feminist scholars like Simone de Beauvoir and Judith Butler. In 1949, de Beauvoir proclaimed in her classic text, *The Second Sex,* that women were not born, but made. Butler echoed this sentiment in her 1990 book, *Gender Trouble,* with the belief that gender is a performance.

The idea that human beings are blank slates upon which gender roles are inscribed has since become universally accepted, a badge of honor reflecting that a person subscribes to the "right" kind of thought. But it is women who will pay the price for this misinformation, because it makes combating discrimination *more* difficult when it does rear its ugly head. If gender is thought to be learned, masculinity will remain the gold standard and femininity will be reduced to aberrations of it. Women will continue to be pressured by society to rid themselves of stereotypically female traits instead of challenging why being a feminine woman is worthy of ridicule.

For individuals who are misogynistic, uncovering evidence that biologically based sex differences exist in the brain will only confirm their biases. They will believe this information was hidden *because* it upholds their views of women being inferior. Hiding the evidence doesn't allow us to argue against those sexist beliefs.

In any discussion revolving around sex differences in the brain, we often hear there has yet to be a scientific consensus on the subject and scientists don't yet know the truth, but also that *no* sex differences can be located anywhere in the brain and that studies showing otherwise have since been disproven, *and* any differences that *do* exist are unduly emphasized, purely the effects of learning and culture, and an example of why "more rigorous" science needs to be done. This odd mishmash of contradictory statements perfectly illustrates how senseless and poorly thought out this line of thinking is.

Scientific studies have confirmed sex differences in the brain that lead to differences in our interests and behavior. These differences are not due to the postnatal environment or societal messaging. Gender is indeed biological and *not* due to socialization.

Biological Sex Differences in the Brain

The first time this issue caught my attention was in 2015. I was nearing the end of my PhD, and on this particular day, a number of my colleagues had been sending out horrified emails about a study that had been published in the *Proceedings of the National Academy of Sciences.* The study claimed that male and female brains were more similar than previously thought: there were no differences in the brain associated with sex; instead, our brains existed along a "mosaic."

If we aren't careful, this will be a step backward for sex research was the gist of what the emails were saying.

The corresponding onslaught of media coverage felt otherwise, hailing the findings as an exciting step forward in dismantling the "myth" that male and female brains were different. Left-leaning journalists were all too happy to disregard the entire body of scientific literature offering evidence to the contrary in the name of one study that was flashy and controversial, but evoking the "good" kind of controversy.

Since then, this study has been refuted by *four* academic studies. In one, a team of sex researchers, led by Marco Del Giudice, an associate professor in psychology at the University of New Mexico, analyzed the exact same brain data from the original study and found that the sex of a given brain could be correctly identified with 69 percent to 77 percent accuracy.[1] That is much higher than what would be expected if our brains were indeed more alike than different.

Of course, differences also exist at the individual level, and this doesn't mean that environment plays no role in shaping us. But, as mentioned in Chapter 1, biological sex dictates gender in more than 99 percent of us. To claim that there are no differences between the sexes when looking at group averages, or that culture has greater influence than biology, simply isn't true. Socialization shapes the extent to which our gender is expressed or suppressed, but it doesn't dictate whether someone will be masculine or feminine, or whether she or he will be gender-conforming or gender-atypical.

Let me explain: Whether a trait is deemed "masculine" or "feminine" is culturally defined, but whether a person gravitates toward traits that are considered masculine or feminine is driven by biology. For example, in the Western world, a shaved head is viewed as masculine, and the majority of people sporting a shaved head are men. For women who choose to shave their head as an expression of who they are, they are likely more masculine than the average woman, and will probably be more male-typical in other areas of their life, too. From a biological standpoint, compared with other women, there's a good chance they were exposed to higher levels of testosterone in utero.

If, in an alternate universe, a shaved head was seen as a feminine trait, we would expect to see the reverse—most people who shaved their head would be *women*, and any men who chose to do so would likely be more feminine than other men, and exposed to *lower* levels of testosterone in the womb.

For someone who is gender-nonconforming, this is similarly influenced by biology, but the extent to which they will feel comfortable expressing their gender nonconformity (through, say, the way they dress or carry themselves) will be influenced by social factors, like parental upbringing and cultural messaging. Societal influence cannot, however, override biology. No matter how much parents or teachers or peers frown upon gender nonconformity (or gender conformity, for that matter), a person will gravitate toward the same interests and behaviors, but he or she may feel more inclined to hide that part of themselves. (I go into greater detail about the effect of nature versus parenting in Chapter 8.)

Activists will also point to intersex people as evidence that biology doesn't always predict one's gender. But because intersex people possess a mix of both female and male characteristics, this can lead to having a gender identity that is different from the way one appears to the outside world. So, even in these cases, biology is still dictating a person's sense of gender. To suggest that this group proves that gender is completely unrelated to biology, or that a person's sense of gender in the brain somehow operates in a way that is distinct from the rest of their body, is flawed and foolish.

Going back to the four new studies that disproved the first erroneous study, care to wager a guess as to what the journalistic response was when they were published? Overwhelming relief and gratitude that scientists had gotten to the heart of the matter? Frantic scrambling to correct their previous reportage, now knowing it was incorrect?

Not quite. There was barely a peep! The vast majority pretended it hadn't even happened. Why threaten social progress in the name of scientific accuracy?

Since then, it has felt like a tedious chore, yanking out the same weed every time a research study pops up denying sex differences in the brain.

Shortly after the "mosaic" study, another found that the sex differences in the size of the hippocampus, responsible for memory, could be explained away by the fact that the male participants in the study were physically bigger (by approximately 10 percent) than the female participants.[2] The size of the hippocampus appeared to scale with men's overall body size.

Feminist neuroscientists have since overzealously tried to apply the same logic to every other part of the brain that's been shown to be sexually dimorphic* in hopes of demonstrating that brain differences aren't due to anything other than men being physically larger.

Shortly thereafter, a study in the journal *NeuroImage* called into question another sexually dimorphic brain region. The amygdala, involved in emotional processing, was shown to be about 10 percent larger in men. But after accounting for men's larger body size, this difference was no longer statistically relevant.[3]

The issue with this study's conclusions, however, is that we don't actually know that a person's body size has any bearing on the size of this part of the brain. By this logic, men who are physically smaller would be expected to have an amygdala that is female-typical, when in actuality, their amygdala would still likely resemble what is typical of men. It's quite possible that a larger amygdala in men is reflective of genuine sex differences in, say, brain function. Even if male and female brains were identical structurally, this doesn't rule out differences in brain functionality.

---

*sexually dimorphic: different between women and men.

Studies have shown that sex differences exist across a wide range of cognitive abilities, including verbal fluency and mental rotation.* In studies using functional magnetic resonance imaging (or fMRI), which measures brain regions that "light up" from changes in neuronal activity during a particular task, women on average outperform men on the former, while men on average outperform women on the latter.[4]

Going back to the amygdala, fMRI studies have indeed shown differences between men and women in this part of the brain when looking at visual sexual stimuli (also known as pornography). Encompassing the emotional component of sexual arousal, the amygdala has a tendency to show greater activation in men than women.[5]

Even if there were no sex differences in the size or function of the amygdala, this doesn't mean there are no differences at all between men and women when it comes to other parts of the brain. For example, the third interstitial nucleus of the anterior hypothalamus, a tiny part of the brain responsible for regulating sexual behavior (at roughly the size of a grain of sand!), has also been shown to be sexually dimorphic, as it is consistently larger in men than women.

Sure enough, when a team of scientists at King's College London published in *Cerebral Cortex* the largest neuroimaging study examining sex differences to date, they found significant brain differences between the sexes, including the amygdala.[6] In a sample including 5,216 brains, the researchers found that even when men's larger overall brain size was

*Verbal fluency refers to the ability to generate many different words beginning with a given letter. Mental rotation refers to the ability to rotate three-dimensional geometric figures in the mind.

taken into account, the amygdala was, on average, larger in men. Another study, published in 2019 in *Nature*'s *Scientific Reports*, found sex differences in gray matter volume among 2,838 participants.[7] Research investigating white matter connections (or connective tissue) in the brain has also demonstrated differences, showing that men had a greater number of white matter connections running from the front to the back of the brain, while women had a greater number of connections running between the two hemispheres.[8]

I can already sense the blood boiling in anyone who believes the sexes must be the same in order to be equal. Frequent criticisms include that authors like me are cherry-picking findings, that neuroplasticity is more important, and that brain-imaging research is unreliable, especially considering a dead salmon once showed brain activation while lying in an MRI scanner.

Regarding the first point, defenders of sex differences are often accused of holding up studies that suit our position and ignoring others that don't. As an impartial onlooker, you don't have to take my—or anyone's—word on what the science says. With the advent of the Internet, you can compare a handful of recent studies claiming some variation of "sex differences don't exist or can be explained away," and *hundreds* of others, including new and old studies, documenting otherwise.

Defenders of the "no sex differences" perspective will also bring up neuroplasticity—the brain's ability to be modified by experience—to support the claim that gendered behavior is not innate. They will treat neuroplasticity as though it is an omnipotent phenomenon capable of overriding all biological influence and sculpting the brain into something entirely unique.

But as Larry Cahill, a neuroscientist and professor at the University of California, Irvine, stated in a seminal 2014 paper in *Cerebrum,* an analogy would be that it's indeed possible to force a child who is naturally left-handed into becoming right-handed, but this doesn't mean it's a good idea, or that she will be capable of using her right hand as effortlessly as her left hand, had it developed without interference.[9] In my opinion, it doesn't make much sense to exploit the plasticity of the brain for the sake of doing so when our underlying predispositions are not harmful, and at the end of the day, they are inescapable.

Neuroimaging is a relatively young field, considering that functional MRI was first implemented in the 1990s, but its methods are rigorous. After a 2009 social psychology study showed that not taking adequate precautions when analyzing one's data could lead to false positives* in, of all things, a salmon's brain,[10] many neuroscientists I knew cringed, but came away with a greater consensus on the types of statistical analyses that should be used. The study was not an attack on fMRI per se, but failure on the part of some scientists to use adequately stringent methods when analyzing brain scans. Nowadays, however, neuroimaging studies are required to meet these statistical criteria before they can be published.

*false positives: refers to regions of the brain that appeared to be activated when they weren't.

If critics want to pick apart a study badly enough, they can find a way to. If they want to hold up one that was poorly done but backs their claims, they can easily do that, too. Many of those most aggressively advocating for the social constructionist view have no training in neuroscience or sexology, but because these opinions are socially pleasing and palatable, they will be accepted without a second thought.

The success of the feminist movement does not require the claim that men and women are identical. Defending this point in the face of contradictory evidence only serves to diminish feminism's cause. When you start denying basic biology, people stop listening, and the larger point about equality is lost.

Many progressives are vocal supporters of science and evolution until it has anything to do with the brain. In their view, evolution stops at the neck in a sort of neuro-creationism, a secular version of intelligent design.

As a liberal, I often find myself in awe at the mental gymnastics required to make sense of such incoherence. Proponents will acknowledge differences between women and men with regard to reproductive anatomy and secondary sex characteristics, yet for some reason, they believe the organ responsible for programming these differences is the same between the sexes?

Neuroscientists pushing this agenda don't seem to have fully thought the consequences through. One of the major implications of male and female brains being the same is that sex won't be considered a relevant variable when conducting research. As a result, findings that are extrapolated from the brain scans of men may or may not have any relevance to women.

For instance, when doing neuroimaging studies pertaining to human sexuality, one important consideration when including female participants is where they are in their menstrual cycle. If a woman is ovulating (and thereby at peak fertility in her cycle), this will influence her sexual response, compared with when she isn't ovulating. Depending on the particular research hypothesis and whether the influence of ovulation on behavior and brain activation is of interest, scientists will schedule a female participant's brain scan around the days of her cycle.

As a result, conclusions drawn from studies using only men won't be applicable to female populations, seeing as how male sexuality doesn't vary according to a twenty-eight-day cycle, and vice versa. Studies using only women won't tell us much about male sexuality. Additionally, research using a combination of men and women grouped together as one population of interest, instead of analyzing them separately as would be appropriate, will lead to muddy results and conclusions that don't really apply to anyone.

Researchers who know a thing or two about sex differences in sexual behavior will be happy to stick to using only male participants because they won't have that pesky thing called ovulation delaying scheduling and holding up data collection. Funding agencies won't require studies to specifically look at the female aspect of any research question because the assumption will be that it doesn't matter. This will lead to the erasure of scientific findings that actually apply to women.

Yet we are able to talk about sex differences pertaining to other parts of the body. For example, women face a greater risk for stroke, while men have a greater risk for heart disease. From a psychological perspective, men are three times more likely than women[11] to be diagnosed with autism spectrum disorder.* But we don't see feminist academics demanding that women be on par with men for risk of heart disease or autism (or that these conditions be given greater attention so that the associated sex differences will lessen), because that doesn't fit their priorities.

There is no reason why society can't take a reasonable approach to discussing sex differences as they pertain to the brain. The issue is not that people are incapable of it; it's that we aren't even having the conversation.

Prior to studying sexology, I remember my reaction whenever I'd come across terms like "neurosexism" and "neurotrash," frequently lobbed as an insult by neuroscientists claiming there are no sex differences, at those who beg to differ—I felt relief. As mentioned, I was once a staunch feminist who promoted the idea that gender was a social construct, and was pleased to see outspoken women in the field shutting down regressive beliefs. I was, in fact, a fan of some of these scholars' work, and to witness several of them calling me out publicly since I've become a journalist has been surreal, to say the least.[12]

My attitudes have changed because I realized that acknowledging

*This is due to what's been called the "extreme male brain." See Baron-Cohen, S. (2002). The extreme male brain theory of autism. *TRENDS in Cognitive Sciences, 6,* 248–254.

these differences and wanting to understand them is not, by definition, a sexist endeavor. When I began reading sexological papers, the realization that female and male sexual systems were at all different completely upended my worldview. I would subsequently go into the lab each day to work and saw that, instead of plotting male domination, my colleagues were merely interested in doing good research. The public, more broadly, must have faith that researchers studying sex differences, and particularly those in the brain, are not doing so as a way to hold back women.

In reality, even if neuroscience showed that every single part of the brain was identical between women and men, this still wouldn't end sexism, because sexist people don't care what the neuroscientific research says. Men who think women are less competent will continue to think so, even if women's brains were carbon copies of theirs. Considering that brain imaging studies cost upward of tens of thousands of dollars, a more effective use of these resources would be to promote a healthy acceptance of these facts while emphasizing that we should treat people as individuals.

It isn't sexist to acknowledge sex differences. What's sexist is assuming that women must be the same as men in order to be treated as equals.

The Google Memo

So, if these differences aren't due to societal influence or stereotypes, how does denying that manifest in the world around us? In August

2017, this very question made international headlines when James Damore's infamous Google manifesto was leaked to the public. The memo illustrated just how contentious the nature-versus-nurture debate had become, and how any attempt at a civilized conversation about gender, sexism in tech, or political censorship would inevitably dissolve into a screaming match of allegations that the other side consisted of alt-right Nazis or science-denying "snowflakes."

At the time, Damore was a senior software engineer at Google, and his memo was feedback invited by the company after he attended a "Diversity and Inclusion Summit" at their Mountain View, California, campus several months earlier. The resulting ten-page document was carefully written, basing its conclusions on studies demonstrating that men and women, on average, find different types of occupations interesting, that these interests have biological underpinnings, and perhaps this was the reason for the gender hiring gap, as opposed to discrimination or bias. The very first sentence read, "I value diversity and inclusion, and am not denying that sexism exists."[13]

From a scientific perspective, Damore's memo was solid. Citing scientific studies, Damore wrote that differences in exposure to prenatal testosterone led women, on average, to prefer people-oriented jobs and maintaining a work-life balance, while men, on average, gravitated toward thing-oriented, high-status occupations. This, of course, wasn't to say that some women weren't more like men, or vice versa, or that all women are one way and men are another—as Damore wrote, "[There's] significant overlap between men and women, so you can't say anything about an individual given these population-level distributions."

But when the memo was published, few appreciated Damore's honesty. Danielle Brown, Google's newly appointed vice president for diversity, integrity, and governance, said the memo advanced "incorrect assumptions about gender."[14] Liberal journalists called the memo an "anti-diversity screed"[15] and Damore, an "alt-right"[16] "tech bro."[17] The citations Damore had provided, linking to research backing his claims, had also been stripped from the document.

Considering that sex differences have been used in the past to hold women back, it's perhaps understandable why some responded in this way. Additionally, Damore's use of technical language—for example, writing that women are higher in "neuroticism," a clinical term referring to one's disposition toward negative mood—would be considered the norm within fields like neuroscience and evolutionary psychology, but outside of this context, probably read as insensitive. In attempting to rectify this and any possibility that Damore's memo would be used to justify discriminating against women in the sciences, many journalists sought to dismiss its contents in their entirety.

You may remember a similar catastrophe unfolding in 2005 around comments made by Lawrence Summers, who was, at the time, the president of Harvard University. In a private conference hosted by the National Bureau of Economic Research, Summers commented on the shortage of women in engineering and science, suggesting that we take into consideration innate causes as opposed to social explanations.[18] In return, Summers was forced to repent and apologize repeatedly. He resigned from his position as university president the following year.

Not much has changed since 2005. After Damore's memo hit the press and people began losing their minds, he was fired from his job for "perpetuating gender stereotypes."[19] He, in turn, filed a class action lawsuit against Google in conjunction with another former Google engineer, David Gudeman, citing discrimination against conservatives and employees who were white or Asian men. Both plaintiffs have since moved into arbitration, but the case continues on behalf of job applicants who believe they were discriminated against.[20]

I defended Damore's memo in a column I wrote for the *Globe and Mail*.[21] When the news first broke, I watched the headlines appear, one after the other, in rapid succession. Reminded of my colleagues' distraught emails about the "no sex differences in the brain" study, I felt so disappointed that the issue was being covered, once again, in such a biased and completely unscientific way.

The column went viral. Within an hour of being published, I was flooded on social media with accusations that I had sold women out, that I was a peddler of pseudoscience and eugenics, and that I was trying hard to be a "cool girl." According to these folks, my opinion was just what you'd expect from someone who's written for men's magazines. Interestingly, many of these insults came from white, male journalists and tenured female professors who would normally be proselytizing the importance of "lifting up" the "marginalized voice" of a minority woman.

I had left the direct messaging option open for my accounts on these platforms, because the week prior, I was crowdsourcing topics of interest from my readers for a column I was writing about kinky sex. Although the majority of public tweets and posts I received were hostile and insulting, quite literally *all* of the private messages voiced support and thanked me. Some were scientists and academics. Others were men—and women—who worked in tech. Most told me they couldn't say aloud the things I had written without fear of losing their job or alienating their friends and family.

The whole experience spoke volumes about social media optics. Those attacking me publicly were doing it to show off to their friends. Whether or not they actually believed what they were saying about me and the science I was defending was another issue altogether.

Claire Lehmann, the founder and editor in chief of *Quillette*, reached out to me for my comments. She published them, along with commentary from three scientists (Lee Jussim at Rutgers University, Geoffrey Miller at the University of Mexico, and David Schmitt at Bradley University) on *Quillette*'s website.[22] Within a day, *Quillette* suffered not one but two distributed denial of service (DDoS) attacks. It was baffling and unnerving, the extent to which strangers on the Internet were willing to go in order to stifle alternative perspectives.

As despised as evolutionary biology may be, it is a critical piece of the discussion on gender. Case in point: I will often hear women who aren't in favor of biological explanations, saying things like "I've never been a girly-girl; I've always been more interested in things that men do," and they will take this as evidence that femininity is the result of an individual's life choices. In reality, from a scientific perspective, women who are gender-atypical, like myself, were likely exposed to higher levels of testosterone in utero. This can occur due to a variety of factors, including normal variation, young maternal age, maternal weight gain, genetic conditions,[23] and hormonal treatment during pregnancy.[24]

As Damore mentioned in his memo, gendered interests are predicted by this exposure—higher levels are associated with male-typical interests and behaviors, regardless of whether the baby is male or female. These interests include a preference for mechanically interesting objects and systemizing occupations in adulthood. Lower levels are associated with a preference for people-oriented activities and occupations, stemming from evolutionary roots. Women, who are tasked with the role of bearing children, evolved to be more sociable, empathic, and people-focused, while men, as hunter-gatherers, were rewarded for strong visuospatial skills and the ability to build and use tools. This explains why STEM (science, technology, engineering, and mathematics) fields tend to be dominated by men.

A big part of the problem seemed to be that many of Damore's critics didn't understand the concept of statistical averages. Even as I summarize the research now, I want to be very clear. Group averages don't say anything about an individual person, but rather, what would be found when comparing a random selection from one group (say, men) to a random selection of another group (in this case, women). More important, Damore was referring to *interest* in scientific fields, not ability.

We see further evidence for this in young girls with congenital adrenal hyperplasia, who are exposed to unusually high levels of testosterone in the womb. When they are born, they prefer male-typical toys, like trucks, even if their parents give them more positive feedback for playing with female-typical toys, like dolls.

Cross-cultural research offers even more evidence that the smaller number of women in STEM is due to women's own choices, as opposed to sexism. Cultures with greater gender equity have larger sex differences in occupational preferences because in these societies, people are free to choose their jobs based on what they enjoy as opposed to what's merely available to them. This trend is apparent in personality differences, as well. In countries with greater gender equity, women remain higher, on average, in traits like neuroticism and agreeableness. (As mentioned earlier, neuroticism refers to one's inclination toward negative mood.) Men, on the other hand, tend to be higher, on average, in traits like narcissism and psychopathy.[25] As gender equity continues to improve in developing societies, we should expect to see these gaps between men and women *widen*.

Another criticism was that coding is a "language," and if the fairer sex excelled at verbal skills, they should be just as proficient at computer programming. I can tell you, as someone who coded during her PhD, that it is not an enviable task. Anyone who thinks programming is glamorous and fun has never spent a serious amount of time doing it. Even if you are sitting in a room filled with people who are similarly coding (and presumably having fun), you are focused on the task in front of you. Those who are good at it don't tend to talk much because it's distracting and you won't get much work done.

To the average person, this sounds more like a nightmare than an ideal situation. Most people, regardless of sex, do not want to sit by themselves for days on end, staring at a computer screen filled with nonsensical words. Only a very specific type of person will find this kind of work enjoyable, and fewer of these people will be women.

Critics were also quick to point out that coding was once dominated by women, and that the advent of home computers, which were marketed solely toward boys, was what led to the sex ratio inverting. As reported on one popular episode of NPR's *Planet Money,* not only were boys getting a head start on their programming skills, but sexism and pushback from professors and male students were also believed to have caused the dwindling of women's enrollment in college computer science courses.[26]

I would argue, however, that a young woman interested in pursuing computer science would do so, regardless of the obstacles in her way. Even in the face of adversity, if a woman wants to succeed, she can. As one of the women interviewed in the episode explained, she switched majors in college, doubting her abilities as a computer scientist, only to later return to the discipline to complete her PhD. Role model theory suggests that it's easier for women to take up a discipline if they have female role models for inspiration and encouragement. While I don't disagree that having female mentors can help, a lack thereof cannot fully account for the paucity of women in STEM.

Jussim, my colleague who was also quoted in the *Quillette* piece, has argued convincingly that the gap in STEM probably has more to do with the fact that girls who possess strong math skills tend to also have strong language skills. In contrast, boys are more than twice as likely to possess strong math skills but *not* strong language skills. In a blog for *Psychology Today,* one of the studies Jussim cited showed that individuals possessing a combination of strong math and verbal skills were *less* likely to pursue STEM jobs than those who only possessed high math skills. Consequently, it isn't a lack of ability that leads girls to go into non-STEM careers, but the fact they have more vocational opportunities available to them.[27]

Countless programs encourage girls and young women to pursue careers in STEM, along with incentives specifically being implemented in an attempt to fix the presumed problem. In August 2019, the University of Technology Sydney announced it would be giving extra points to the university admissions test scores of women applying to undergraduate STEM degrees. The goal was to increase the number of women working in engineering, construction, and IT sectors.

Although probably well intentioned, these kinds of policies are not ultimately beneficial because they suggest that female students can't attain the cutoff scores on their own merit. They also don't help to combat the myth that women aren't as inherently capable. Imagine how women in STEM programs will be treated if the perception is that they were only admitted because of systematic handouts. In order to maintain a 50:50 sex ratio, we will be forcing women to do jobs they don't want to do.

Damore has since publicly disclosed that he is on the autism spectrum.[28] Looking back, I don't believe most journalists were in fact ignorant about the truth about gender, sex differences, or prenatal testosterone, and I find it disturbing how willing they were to sacrifice Damore—and the truth—in the name of creating controversy and news bait. Those who are misogynistic will only see this as further evidence that (a) the media can't be trusted; and (b) women don't know what they're talking about.

As for the future of neuroscience—and scientific discovery more widely—as a greater number of public and private companies and scientific institutions jump on the social justice bandwagon, we will be hard-pressed to find scientists publishing papers demonstrating sex differences. Any such differences that are biologically based will be reframed as the result of socialization or cultural influences, not sex (as discussed in Chapter 9). It will create a chill among research experts, discouraging them from speaking publicly because no one wants to be called sexist.

Not wanting to be all gloom-and-doom, here is one potential upside: not everyone currently going along with the blank slate narrative necessarily believes it. Many will privately acknowledge that brain-based sex differences do exist, and this is all a charade in the name of helping women move forward. They fear that admitting that sex differences exist will justify female oppression.

In theory, the separation of gender from biology and sex would help to further facilitate female emancipation. If gender and gender roles are socially imposed, it becomes more difficult for society to justify defining women and their opportunities by these roles, instead of encouraging them to flourish as individuals.

The bottom line is, all of this messaging *sounds nice*. I can see why people like it. It absolves us of having to wrangle with the complexity and the implications that come from acknowledging that our gender is mostly hardwired and unchangeable. If gender is learned, you can change it. If there's anything you don't like about who you are, it's not your fault; it's the fault of cultural conditioning.

If women aren't choosing occupations in science and tech because they'd prefer to fulfill other career aspirations, this shouldn't be interpreted as a problem. As a woman with a PhD in a STEM discipline who has mentored female students, I understand why this perspective is seen as dangerous. It could be used to drag women back a century or two, to deny our basic freedoms. But fearmongering isn't honest, ethical, or empowering. Women should be allowed to own the decisions we make. There is nothing more infantilizing than the belief that women who think independently aren't aware of their own oppression.

Sexism in STEM

As for the question of whether STEM fields are cesspools of misogyny, a report published by the U.S. National Academies of Sciences, Engineering, and Medicine suggested that as many as 58 percent of women in academia have experienced sexual harassment, including 43 percent of graduate students in STEM. Female graduate students were also 1.64 times more likely than male students to be on the receiving end of sexual harassment from faculty and staff.[29]

Sexual harassment can be experienced by anyone, and even those working for world-class scientists at prestigious institutions aren't immune. I, too, experienced sexism when I was working in research and have seen the depths of the abyss. Of some of the more family-friendly examples I can share in this book, one of my superiors once called me a bitch to my face. Another time, a colleague felt the need to tell me that if my breasts were bigger, I could forget about the whole "grad school thing" and pursue a more lucrative career doing porn.

These experiences were in addition to many other instances of garden-variety sexism that every woman faces, like when senior male researchers would speak in a patronizing way that never seemed to manifest when they were talking to less experienced male students.

Sexual harassment is unacceptable and women shouldn't have to tolerate unwanted or discriminatory behavior. Needless to say, harassment can have severely negative effects on a person's psychological and physical health, beyond affecting their work, morale, and productivity.

But just because women—and men—have had to endure experiences like these or worse, this is not a reason to distort what the science says, or to blame men for the shortage of women in STEM. In some cases, academic policies actually favor the hiring of women over men.

One study from 2015 showed that women are more likely to be hired as faculty in STEM disciplines, with a 2:1 preference for qualified females over equally qualified male applicants.[30] In another study,[31] the same research team conducted hundreds of analyses and similarly determined that women in the academic sciences—ranging from the life sciences and social sciences to mathematically intensive fields like computer science and engineering—were more likely to receive hiring offers than men. Sexism also appeared to be less of a factor when it came to female academics' professional advancement—they received tenure at the same rate as their male counterparts, their funding applications were equally accepted, and their publications were equally cited.

Ignoring these data discounting sexism as the sole explanation for the gender gap in STEM doesn't help to strengthen the argument for gender equality. What it does instead is amplify resentment in those who are aggrieved that women are taking up space in these disciplines, because only presenting one side of the argument comes across as being biased and agenda-driven. A more helpful conversation would emerge from taking both perspectives into account—yes, sexism exists, but there are also initiatives countering it—while encouraging young women to be assertive about what is and isn't acceptable behavior. Discussing practical issues, like how to deal with threats of retribution, are more productive and helpful than terrifying women into avoiding the field altogether.

On the flip side, writing about this issue has been enlightening for me, and reminds me why so many have embraced the false "gender is a social construct" explanation wholeheartedly. I have come across plenty of men who are walking, breathing explanations as to why defenders of sex differences receive so much pushback for things we say. I feel a similar sense of disgust and repulsion whenever I come across fringe lunatics who cite sex differences as evidence that women can't be scientists and who criticize women for working outside of the home instead of prioritizing their fertility and sexual attractiveness to men. These individuals serve as a personal reminder to be thoughtful and responsible when discussing this issue, because the last thing I want to do is give ammunition to the other side, to those who *do* view women as subordinate.

There exists a middle ground between taking this information and running with it in either direction. I've found that the individuals who most vocally criticize the underrepresentation of women in STEM, attributing this disparity to toxicity in the workplace, are often not scientists themselves. They are usually graduates from disciplines like gender studies or education, who have no formal training in the sciences, no knowledge of how the scientific method works, and no actual experience in the field.

I was fortunate to grow up in a household in which I was encouraged to pursue whatever I wanted. I was never told that I couldn't do something because I was female. Perhaps that is why I don't feel I have a stake in this debate besides having been a scientist who took a lot of pride in her work. I don't see sex differences as inherently threatening, nor do they inspire any animosity or rage within me. I am sympathetic to those who have had negative experiences. We should want to live in a world where women and girls are not punished because of their sex, but it isn't necessary to hold science hostage in order to do so.

At its outset, feminism had a laudable goal: equal rights for women in the realms of educational and occupational opportunity, including higher education, working outside of the home, owning property, and voting. But its core concerns have shifted more recently to promote a false premise—that as a society, we have not reached gender equality until men and women are identical. It is a claim that is not only specious, but one that will ultimately harm women's ability to make autonomous and enlightened choices. The current disconnect between feminism and biological science is unnecessary, because there is no reason why scientific evidence can't be used to further feminist goals.

Instead of moving forward in this way, the corresponding response has been to embrace a wealth of misinformation and to water down the definition of what it means to be a woman so that we will become socially acceptable. Young feminists should be fighting for the right to be respected as women, whether we are female-typical or atypical, instead of trying with all their might to win someone else's game.

MYTH #3

THERE ARE MORE THAN TWO GENDERS

There are only two genders. There. I said it.

Not three, not seventy-one, and certainly not an infinite number. Gender is not a spectrum, a continuum, a kaleidoscope, a prism, or any other majestic-sounded metaphor that gender activism has dreamed up.

With the number of genders increasing exponentially by the day, it's hard to stay on top of things. You've likely heard that some people identify as both genders or neither, and that others have a gender that alternates from when they wake up until they go to sleep. One BBC film used to educate schoolchildren during health class suggested there are more than 100.

No. There are two: female and male. There is zero scientific evidence to suggest that any other genders exist.

The Terminology

The belief that gender is a spectrum has been promoted by mainstream media, scientific journals, and medical organizations alike. In the last few years alone, countless celebrities have "come out" as nonbinary. The *Canadian Medical Association Journal* published a commentary that referred to gender identity as a "spectrum" or "galaxy," as opposed to being binary.[1] *Nature*, one of the most prestigious scientific journals in the world, ran an editorial referring to individuals who exist outside the male-female binary.

A multitude of social and policy changes have also been implemented. The Equal Employment Opportunity Commission created guidelines for employers on how to report nonbinary employees so that they can be accounted for in the workplace. One state after another has begun issuing gender-neutral and nonbinary documentation, using a designation of "X" in place of male or female, based solely on self-identification. The gender-neutral title "Mx" is increasingly being used to address nonbinary people. Merriam-Webster's dictionary added a new entry for use of the singular pronoun "they" to refer to a person whose gender is nonbinary. Not to be outdone, a London zoo announced it would be raising a genderless penguin.

People who identify as *gender nonbinary*, or *enby* for short, don't identify exclusively as either male or female.* Nonbinary identity labels may seem boundless, but they all basically mean that a person identifies, to some extent, as both male and female, or neither. *Genderqueer,* for example, means a person identifies as neither, both, or a combination therein. *Bigender* and *androgyne* refer to someone who identifies as both, *trigender* refers to someone who identifies as exactly three genders at the same time (like female, male, and nonbinary), *quadgender* refers to someone who identifies as *four* genders at the same time (we'll get to additional possibilities in a second), and *pangender* refers to someone who identifies as "all" genders. For the average person, this can all be a tad overwhelming and confusing.

Contrasting with this are terms like *gender-neutral, genderfree, agender,* and *neutrois,* which all mean that a person doesn't have a gender. *Aliagender* means a person identifies as a gender that is not male or female.

Then there is a family of identity labels that refer to the degree to which a person identifies as one gender slightly more than another. *Demiboy, demiguy,* and *demiman* refer to someone who identifies as partly a boy and partly another gender, regardless of their birth sex. *Demigirl* and *demiwoman* similarly refer to someone who identifies partially as female. An individual who identifies as *feminine-of-center* feels feminine, but doesn't identify as a woman, just as someone who is *masculine-of-center* feels masculine but doesn't identify as a man.

*enby: a phonetic pronunciation of the letters "N" and "B," derived from the first two syllables of the word "nonbinary."

Terms like *girlflux* and *paragirl* (or *boyflux* and *paraboy*) mean that a person identifies mostly as that gender, but not fully. *Librafeminine* (or *libramasculine*) means you identify mostly as neither gender, but are at the same time a bit female or male.

Finally, there is *genderfluid*, which means a person's gender changes every so often, whether it's by the hour or by the day. *Amalgagender* means a person's intersex condition is a part of their gender. *Anongender* is someone who doesn't know what their gender is.

Some will also use initialisms like "FTX," which describes someone who was born female and now identifies as nonbinary. (Similarly, "MTX" describes someone who has gone from being male to "enby.") Stylistically, this takes after transgender nomenclature, in the way that FtM signifies someone who was born female and identifies as male.

In researching this terminology, my favorites have been *moongender*, which means your gender comes out only at night; *puzzlegender*, which means you have a gender that feels like it needs to be pieced together; and *arborgender*, which means you identify as a tree.

If you can believe it, this isn't an exhaustive list. There is also an entire lexicon of words to describe discrimination against nonbinary people, like *enbiphobia* and *femmephobia.* I learned that *gender-variant,* which was considered a progressive way to describe gender-nonconforming people only a few years ago, is now seen as "problematic" because it implies that nonbinary genders are deviations from the two "normal" genders.

In my estimation, of all the different labels, the four most commonly recognized ones are nonbinary, genderqueer, gender-neutral, and genderfluid. Sometimes these labels incorporate the "assigned at birth" language discussed in Chapter 1, like "DFAB nonbinary trans femme," and "AMAB genderqueer girl."

As you know from Chapter 1, humans are a sexually dimorphic species, with two types of gametes: eggs and sperm. Intermediate gametes don't exist. Since biological sex and gender are both defined by these parameters, gender is, by definition, like sex—either male or female; binary and not a spectrum.

Everyone, to some extent, is a combination of male and female traits. No one is 100 percent male or 100 percent female or 100 percent gender-conforming. When we think of the average man, he's probably interested in a variety of things that most guys are into, like playing sports and building things, but maybe he also enjoys watching talk shows and talking on the phone. By the same token, there are plenty of women who like playing video games and competing in sports. (These are, by the way, gendered traits commonly used in social psychology research, so I can't be called sexist for using them as examples.)

It's not appropriate to consider everyone who exhibits some degree of gender nonconformity as belonging to a third category of gender. It's more outdated to assume that someone who is gender-atypical is another category of gender entirely, than to consider them as part of the normal variation you'd expect within female or male, just like any other human trait.

Gender is not a continuum or a rainbow or a diverse spectrum. It exists as two discrete categories, female and male, not as two polarities along a shared continuum along which human beings appear with equal likelihood. Even if we were to divorce gender from sex, the vast majority of people fall clearly into one of these two categories, as opposed to being equally spread somewhere between them.

I'm also not sure how an entire social movement has managed to overlook this, but if a person says that they are both genders or neither, this still depends on the concept of gender being binary. Even for those who identify as gender-nonconforming, it means they are less like one gender and more like the other.

It won't surprise too many of you that most of those traveling this new gender topography are millennials and Generation Z. According to a Harris poll conducted on behalf of GLAAD, 10 percent of millennials self-identified as transgender, agender, genderfluid, bigender, or genderqueer, compared with only 4 percent of Generation X who said the same.[2] Greater awareness and social acceptance of having a transgender identity surely played a role in this increase, but my sense is it goes beyond that.

From what I gather, being nonbinary is a way to experiment with self-expression, the way that everyone did as teenagers. Millennials in particular get a bad rap for being self-centered, dubbed the "Me Me Me generation," so for detractors, these findings would seem apropos. In decades prior, carving out one's identity came in the form of being skateboarders or goth, or in my case, punk rock. Some of us got body piercings, tattoos, or poorly thought out haircuts, in addition to making questionable fashion choices. That was part of the beauty of being young and figuring out who you were. But adults recognized it for what it was, and we were never authoritarian about it.

That brings us to the word "queer," which is heavily loaded for a number of reasons. It is defined by GLAAD as "experiencing sexual attraction in a way that does not fit into . . . dominant norms." The word originated as a slur against people who are gay; some gay people have reclaimed the word, but for others, it remains a term of abuse. Within progressive circles, many otherwise ordinary, straight people have taken ownership of the word, because identifying as "queer" grants them the ability to claim sexual minority status. In a new study from the Williams Institute at the UCLA School of Law, the majority of people identifying as "queer" are nontransgender women. In my opinion, for straight men, identifying as "queer" can be a tool signaling their progressivism so that feminist women will date them.

You may wonder why a group of otherwise straight, female-typical women would want to identify as part of an oppressed class. The answer is intersectionality, a term coined by Kimberlé Crenshaw, an American civil rights advocate, thirty years ago. At its crux is the belief that women and minorities experience systemic injustice, and that those who are not members of minority groups are considered to have privilege. For women who are also racial minorities, the discrimination they face is considered unique, distinct from the discrimination white women and male racial minorities experience.

At its core, intersectionality has some validity in describing how discrimination in society differentially affects women, nonwhite people, and other minority groups. Mainstream feminism, however, has encouraged an entirely new, malignant application, with groups who would otherwise be considered to have privilege clamoring for minority status so that their opinions are given more weight.

Seizing one, or a handful, of the gender nonbinary identity labels allows a person to join the LGBT+ movement tangentially, without question, even when there is nothing L, G, B, or T about them. (As for the term, "LGBT+," it is made up of distinct factions that don't really have much to do with each other.) For instance, bisexuality has been redefined to include people who have romantic feelings for people of both sexes. The "sex" part of "bisexual" is no longer required. Doing so lowers the bar for people to self-identify their way into the community.

From what I've observed, sexual orientation does, however, play a role in the decision to identify as nonbinary. Because lesbian women tend to be more male-typical in some respects and because they are sexually attracted to women, some gravitate toward the nonbinary label because it is more socially acceptable than being a masculine woman or identifying as gay. (The same applies to gay men, which I will discuss in a minute.)

A big part of the nonbinary movement is the refusal to subscribe to traditional gender norms while also denying that gender has any meaningful correlation to biological sex. What it fails to realize is that it's possible for a person to be gender-nonconforming while identifying as simply a woman or a man.

By nonbinary activists' definition, everyone on planet earth is gender nonbinary. If a man decides he sometimes enjoys putting on women's clothing, does that mean he's actually genderfluid? Not necessarily. He may just be a man who, for whatever reason (as discussed in Chapter 4), likes to wear women's clothing.

And to be clear, saying that gender is binary is not the same thing as saying that prejudice against gender-nonconforming individuals is acceptable or that we should impose gender conformity on them. We can acknowledge that some people are gender-nonconforming or gender-diverse, and encourage them to be who they are, without rewriting science to facilitate this acceptance.

Third-Gender Pronouns and the Transgender Community

In a recent Pew Research Center study, roughly one-third of Gen Z and one quarter of millennials reported knowing someone who uses nonbinary pronouns like "they," compared with only about one-sixth of people in Generation X, with the proportion continuing to drop for older generations.[3]

On the Democratic campaign trail for the 2020 United States presidential election, pronouns were a hot talking point. Presidential candidates Elizabeth Warren, Julián Castro, Pete Buttigieg, Cory Booker, Tom Steyer, and Kamala Harris updated their Twitter bios to include their pronouns (she/her and he/him). During a CNN town hall meeting, Harris additionally let the audience know that her pronouns were "she/her/hers" before answering their questions.

Along with he/him/his and she/her/hers, there are also pronouns of the gender-neutral variety. As mentioned, Merriam-Webster acknowledged they/them/theirs pronouns in its recent entry. There is also zie/zir/zirs and sie/hir/hirs and ey/em/eirs.

As critical as I am of the nonbinary movement, I do think language matters, and I will use the pronouns someone wants me to use. But sometimes I get the sense that the language thing has more to do with controlling others than the desire for people to be respectful. Part of the mandate when advocating for this movement is dogmatism about self-determination.

With the growing popularity of gender fluidity and gender neutrality, more people are deciding that announcing one's pronouns at the beginning of any social interaction is a good idea. This has pitted the nonbinary community and its vocal allies against some members of the transgender community, who do not appreciate the sentiment.

Transgender people transition because they want to live as the opposite sex. To ask a trans person explicitly about their pronouns can be offensive, because it suggests their gender wasn't obvious. When non-trans people put their pronouns in their bios and email signatures, et cetera, in an attempt to be more inclusive, this can make it harder for trans people to find other members of their community. Although progressive theories about gender lump nonbinary people under the trans "umbrella," trans people and nonbinary people are not one and the same. Some people identify as both, but some trans people, particularly those who have undergone medical transition, feel they lack any commonality with nonbinary people, and do not wish for the gender binary to be removed.

This is illustrated in one scene from a documentary titled *Gender Revolution,* by *National Geographic*, in which journalist Katie Couric sits down with two trans women—Renée Richards, the first transgender woman to play tennis in the U.S. Open after transitioning in 1975, and Hari Nef, an actor on the television show *Transparent.* At one point in the interview, Nef, twenty-seven, who considers gender to be fluid, describes male and female as "just wisps of smoke."

Richards, eighty-five, acknowledges that she lived forty years as a man and is now enjoying forty years living as a woman, "but that doesn't mean that I'm genderfluid." She adds, "The idea of a binary is what I think the world is. It's the spice of life. . . . It's appealing and I like it," to which Couric muses, "Maybe it's just a generational divide."

I don't think the heart of these two perspectives must necessarily be in opposition. We can accept that gender is binary without boxing people into preconceived gender roles or stereotypes. At the same time, claiming that anyone identifying however they like is *no different* from someone who has medically transitioned within the gender binary (from male to female or vice versa) leads to confusion about gender dysphoria and the inability of medical professionals to help those who are suffering. This is because the root causes of these two presentations are very different.

Indeed, within the conversation about transgender rights has emerged a debate about whether nonbinary people should be considered transgender. Over time, concerns about nonbinary rights have begun to dominate this discussion in online spaces and within the community. For those in support of nonbinary rights, the belief that someone must experience gender dysphoria and undergo medical transitioning in order to identify as transgender is seen as exclusionary because it requires a certain bar to be cleared in order for an individual to be part of the community. To question whether nonbinary people are the same as trans people is derisively known as "transmedicalism."

A few thoughts: as unexpected and difficult as it may be to comprehend the nonbinary movement, I believe it's important to be compassionate because in many cases, an individual who identifies this way is communicating that they are experiencing distress and discomfort. In some cases, a person may legitimately be struggling to figure out their gender, and with that comes much introspection and pain. I don't believe mockery or making fun of nonbinary people will lead to anyone changing their minds, nor does doing so allow for honest dialogue to unfold. My issue is that none of these ideas are backed by science, and that pretending they are only directs the focus away from the underlying issues we should be addressing, issues that I will soon get to.

For those who are skeptical, identifying as a third gender may seem to be a fad. Gender has become trendy and being something other than female or male sounds exotic; a person appears interesting and ahead of the curve. By simply being known as nonbinary or genderqueer or going by "they/them" pronouns, it immediately conveys to other people that this particular individual is *different* and to some degree, special, because they are an exception to the rule, which could be expected in young people and especially teenagers.

As more people take on these labels, being nonbinary has become a way to find community, a sense of belonging, and acceptance. It's not so much about individuality as it is about group membership. In some cases, a person will identify as being a third gender without modifying anything about themselves beyond their clothing and pronouns.

As mentioned, the word "transgender" has expanded to encompass anyone who feels, in any way, different from what would be expected of them, based on their birth sex. This includes gender-atypical and gender-nonconforming people and anyone who feels even mild discomfort about their bodies (which I will elaborate on). The widening of the application of what it means to be transgender means more people will potentially identify this way, thereby inflating its prevalence in the general population. If a larger percentage of people identify as transgender or nonbinary, this offers support for the argument that these identities are real phenomena and discrimination against them is unjust. But we can advocate for this acceptance without socially engineering numbers in favor of it. Doing so only leads to a further lack of clarity for those who will be inappropriately grouped as part of the community.

For the purposes of this book, whenever I refer to the transgender community, I am referring to those with gender dysphoria (who identify more with the opposite sex than their birth sex), who are taking steps to transition to the *opposite* sex, whether it is socially or medically.

To collapse all of these labels into one haphazard group that includes nonbinary people *and* transgender people who do *not* have gender dysphoria, and to then call everyone "trans," diminishes the suffering of those who experience gender dysphoria, a legitimate condition that is recognized by medicine and science. In contrast to how those who are genderfluid describe their gender, gender dysphoria is not a whimsical feeling that comes and goes depending on which direction the winds are blowing that day.

If anything, the concept of gender fluidity calls into question the very idea of being transgender. The concept of gender dysphoria rests on the idea that gender is innate and that the brain of one sex exists in the body of the other. Gender fluidity suggests the very antithesis to this, that one's internal sense of gender can vary by the hour. If one's gender can change, why shouldn't it be malleable to align with a person's birth sex? The concept of gender fluidity argues against trans adults' right to transition.

I understand why the nonbinary movement has gained such momentum so quickly and why it's being held in such high regard. Other sexual and gender minority groups, including gay and transgender people, have long had people telling them that what they experience isn't real or that they're just going through a phase. Part of the push to accept people identifying as nonbinary (and also, children who say they are transgender; see Chapter 5) stems from the empathy of those who want to correct for mistakes in the past.

Then we have celebrities who are capitalizing on gender's moment in the sun. If someone has a large public following and they announce they are nonbinary, it is hard to believe that doing so is anything but a publicity stunt to get people talking about them. Identifying as nonbinary provides social, and literal, currency, and public figures who "come out" as nonbinary obtain adulation and a newfound relevance as every media publication in the world hails them for their courage and strength as a trailblazer.

What I find particularly sad is the number of gay men who have embraced the nonbinary label. When asked in media interviews about how they knew they were nonbinary, many profess a love of being a man, but also wearing makeup and high heels. A few prominent media personalities come to mind here, but they shall remain unnamed. One of the bedrocks of the gay community is drag shows, in which adult men dress up as very feminine women to lip-synch to pop songs promoting female independence and empowerment, all while maintaining an in-your-face sense of humor. It's unclear why, for these individuals, being a drag queen on their days off wasn't good enough. In turn, the next generation of young gay men, who look up to them, are identifying as nonbinary instead of gay men. It's also worth mentioning that feminine gay men and masculine lesbians, by virtue of being gender-nonconforming, are not trans or nonbinary; they are gender-nonconforming men and women.

Based on the Harris poll's findings, millennials are almost twice as likely to identify as "LGBTQ" as people in Generation X. Despite this, millennials who are not part of the community were *less* likely to know someone who identified as "gay" or "lesbian." This means those who consider themselves to be "LGBTQ" are identifying as something other than lesbian or gay, speaking to the broadening of identity labels used to describe one's sexual orientation.

One interpretation of these data would be that more people are identifying as part of the community not out of a true sexual identity, but because it is trendy to do so, or because they believe sexuality is "fluid." This undermines biological explanations for sexual orientation, which has been a disturbingly growing phenomenon and one I never thought I'd witness. Gay rights activists fought for the "born this way" narrative because it came with greater social acceptance—and not only that, it is scientifically correct (see Chapter 4). A person's sexual orientation shouldn't have to be hardwired (and thereby, nonvolitional) in order to be accepted, but it doesn't appear that nonbinary activists have thought through the implications of denying biological evidence. If being gay is a choice, it makes it more difficult to advocate against attempts to change it.

Some nonbinary people will say that they don't like labels like "gay" or "lesbian" because they are binary terms; attraction to a man or a woman, even in the context of being gay, is still considered attraction to a binary gender. Others have argued that you can be a "nonbinary lesbian," which doesn't make sense. In order to be a lesbian, by definition, one needs to be a woman. The whole thing illustrates how ill-conceived these theories pertaining to the gender spectrum are.

Third-Gender Cultures

The transgender community, as well as intersex people, are commonly cited as examples when individuals argue in favor of conceptualizing both gender and sex as a spectrum. But pointing to transgender and intersex people actually offers support for, instead of against, the gender binary. The term "transgender," by definition, means that a person identifies more as the opposite sex than their birth sex, which still operates within a framework of gender being binary. As mentioned earlier, the word is being redefined, as part of this obliteration of the binary, to include identifying as a third gender.

As for intersex people, many are content to live as one or the other, male or female. It just happens, in some cases, that the sex they'd prefer to live as was not the one listed on their birth certificate. Most intersex people are not seeking to nullify gender or to live in a genderless world.[4]

Those on the side of redefining the gender binary will often cite third-gender cultures as examples of gender fluidity. For example, in Samoa, the *fa'afafine* are individuals who are feminine in appearance, biologically male, and attracted to men. (The word "fa'afafine" translates to "in the manner of women.") These individuals live as another gender between male and female, and in Western culture, they would be considered gay men.

In day-to-day life, the *fa'afafine* wear female-typical garments like dresses and high heels and cosmetics. They help out with cooking, raising the young, and taking care of the elders. Their gender nonconformity is usually present from a young age, as adults in the community are able to identify them early on in life.

It's also important to note that the *fa'afafine* do not consider themselves female, nor does their culture. The *fa'afafine* acknowledge that they are male and have male biology, including a penis. As a result, it isn't appropriate to group them as part of the transgender community because they don't transition to the opposite sex and have no desire to become women. They aren't considered the opposite sex within their culture and do not wish to be. Applying the transgender label to people who live as a third gender in these cultures is reinterpreting their way of living through a Western lens.

The way third-gender individuals in Samoa have been integrated into society is analogous to other cultures in which third-gender people exist. In Oaxaca, Mexico, individuals who are biologically male and sexually attracted to men are known as the *muxes*. *Muxe nguiiu* tend to be masculine in their appearance and behavior, while *muxe gunaa* are feminine. This also encompasses the *hijra* in South Asia and the *kathoey* in Thailand.

Those of us defending biological explanations for gender are sometimes accused of neglecting to take cultural influences into account. In the case of third-gender cultures, context does indeed play a role in how gender is defined, but it doesn't actively influence or redefine it.

Individuals who are considered part of a third gender are classified this way due to biological influence (specifically, prenatal testosterone exposure), *not* the influence of society. We can accept the existence of third-gender cultures while maintaining that gender is binary.

Another example frequently brought up to support the idea of the gender spectrum is the existence of the *guevedoces*. The *guevedoces,* which translates to "penis at twelve," are especially common in a small town in the Dominican Republic. When they are born, they have XY chromosomes and testes that are typical of boys, but they have a female external appearance and so they are raised as girls. But at puberty, they develop into boys.

This is because they have a condition called 5-alpha reductase deficiency. The enzyme 5-alpha-reductase normally converts testosterone into dihydrotestosterone (DHT), which masculinizes the body in the womb and leads to the development of male genitalia. For boys with 5-alpha reductase deficiency, the lack of DHT disrupts this growth. It is only when testosterone is produced by the body at puberty that virilization occurs, leading to the growth of a penis and scrotum, a deepening of the voice, an increase in muscle mass, and hence, the change from female to male. Some *guevedoces* will carry on as men, while others will undergo surgery to remain female.

The greater prevalence in this particular region is due to its remote location and the heritability of the condition. Populations of *guevedoces* have also been documented in countries like Turkey and Egypt.

The *guevedoces* are not a violation of the gender binary, nor are they proof that gender exists as a spectrum. The changes these children undergo pertain to primary and secondary sex characteristics, so they don't really have anything to do with gender. As unique and remarkable an occurrence as this may be, it still operates within the framework of sex and gender being binary. Attempting to prop these children up as evidence to the contrary is yet another case of activists grasping at any possible example to justify a predetermined conclusion.

The Conversation We're Not Having

So, what is the nonbinary trend really about?

One account that stood out to me was James Shupe's story. I first became aware of Shupe when he and I both appeared on a popular science podcast several years ago. At the time, Shupe identified as nonbinary and was taking cross-sex hormones. A year earlier, he had become the first person in the United States to obtain legal recognition of a nonbinary gender, with a sex marker of "unknown" on his birth certificate. This set the precedent for nonbinary laws in more than a dozen states.

Shupe has since returned to identifying as male, stating that his nonbinary identity stemmed from trauma while serving in the U.S. military for almost eighteen years and the fact that he experiences autogynephilia[5] (see Chapter 4). Shupe's initial decision to come out as nonbinary was hailed as groundbreaking by what seemed to be every major news publication at the time. Few outlets have been willing to offer him a platform to discuss his realizations since then.

In doing research for this chapter, I watched many (very many) online videos of people identifying as nonbinary, genderfluid, genderqueer, or some variation thereof, talking about their experiences. I wanted to understand their perspective better. As a scientist by training, the first place I went to was the research literature, but there have yet to be any studies critically examining this phenomenon.

For anyone who hasn't come across this kind of content before, it usually begins with a person introducing themselves with their name, their preferred pronouns, and their chosen identity label, such as "a queer, transmasculine person," "a non-gendered shape-shifting person," or "a queer nonbinary trans femme on the spectrum of asexuality," before getting into the meat-and-bones topic of the video.

What I was looking for was something that threw a wrench into my superficial understanding. In so many videos, nonbinary people referred to those dinosaurs who "don't understand" (aka me) and I sincerely wanted to know what it is I've been missing. When I was younger, I remember arguing with the older generation about gay rights, adamant that my friends deserved the same rights as everyone else, and there was nothing wrong with being gay. Now I find myself on the opposite side of the progressive debate and sometimes I question whether I may be wrong.

The way nonbinary people describe their gender is as though it is a mythical creature. The level of specificity that goes into describing this aspect of themselves is carefully considered and painstakingly thorough. It is like peering into a tesseract that is glinting in the sun; there are layers upon layers and different luminosities to their gender, each of which changes depending on the day and the particular situation.

A common theme I've noticed is wanting to rid oneself from societal expectations around gender. In one video, a genderqueer person explained how identifying this way allowed them to wear spiked hair and dark makeup. Another individual, who identified as nonbinary, said they enjoyed being able to wear whatever they wanted, including differing hemlines of men's shorts.

You might be thinking, isn't that androgyny? Kind of. In today's terminology, androgyny refers more to gender expression, like clothing, while nonbinary refers to identity and how a person feels internally.

What I can't understand is why doing any of these things requires changing your gender identity to something completely new. It's perfectly fine to behave in an atypical way but doing so doesn't require overturning every last social convention. The most regressive view is that anyone who enjoys dressing like the opposite sex or who feels as though they are a mix of both male and female must really be another category altogether.

Another theme that is prominent throughout these videos is a cult-like thinking that extols to viewers the importance of embracing "your truth," and that being curious about nonbinary people probably means you are part of the community, too. By that logic and based on the number of hours I've logged watching videos titled along the lines of "I am nonbinary," "I didn't know I was nonbinary," or "COMING OUT AS NONBINARY," that would mean I must be nonbinary, also.

As a grown woman, I am able to watch these videos while remaining critically minded. But for a young person who is still in the process of figuring out who they are, I imagine it would be confusing, especially for those who are suggestible. The trend seems to have been taken up mostly by men who once identified as gay, as well as women, who experience anxiety and discomfort with their birth sex. Evidently a new era of gender has been created in order to avoid dealing with good ol' sexism and homophobia.

It is a new strain of misogyny, with young women seeking to distance themselves from sexist stereotypes that often go along with being female. Countless nonbinary people, explaining their decision to identify as a third gender category, would explicitly say they didn't like being stereotyped as a woman. One person rationalized that women are weak and they felt strong, so logically, they were more like a man, or somewhere in between, or at least anything but a woman.

Girls who are even slightly masculine, or who prefer to wear men's clothes because they are more comfortable, now believe this makes them something other than a girl. In truth, I'd argue very few women enjoy squeezing into dresses or skirts and having to sit awkwardly in public places so that they don't accidentally flash unsuspecting onlookers. This is why sweatpants were invented.

Many speak about feeling "dysphoric" when they get their periods, to which I wish I could reach through the screen to shake them and say, "NO WOMAN LIKES GETTING HER PERIOD! That is just a normal part of being a woman!" Instead, they are being encouraged to consider themselves as part of a nonwoman class due to a natural, biological function.

One could argue this is just part of growing up, and I would agree if it weren't for the accessibility of surgery. Many prominent nonbinary activists speak openly about undergoing a double mastectomy. It is chilling to what extent surgically removing one's breasts is a critical feature of the culture of this community.

In one case, someone who looked to be in their early twenties spoke about how they came to decide on their nonbinary identity. They had been born male, transitioned to female (including undergoing hormonal interventions and surgery), but then realized they didn't identify with what it meant to be a woman, because, in their words, they didn't feel fully comfortable being around other women. So they decided they must not be a woman, but instead, nonbinary. Based on conversations I've had with detransitioners (see Chapter 5), doing so allows those who have detransitioned to maintain support from the trans community, instead of being excommunicated.

In another case, a young person who was born female transitioned to male (again, undergoing hormones and surgery), but missed wearing makeup. As a result, they came out as nonbinary.

Some identify as a "transgender nonbinary" person, because identifying as nonbinary can involve also identifying as the opposite sex and also transitioning. You might be wondering, if a person is neither male nor female, how can they also transition to the opposite sex? What is the opposite of neither a man nor a woman? In this case, it seems, a person will identify mostly with the sex opposite to their birth sex, but upon transitioning, not quite feel like that encapsulates who they are. They will then switch to identifying as nonbinary as a sort of happy middle ground. In truth, it sounds more like they are experiencing transition regret, but they aren't quite ready to fully go back to identifying as their birth sex. For those who have medically transitioned, detransitioning means they will have to live with the permanent side effects of surgery and/or having taken hormones that will make it more difficult for them to present as their birth sex. For those who only underwent a social transition, detransitioning requires acknowledging to everyone in their life, including themselves, that they made a mistake, and as alluded to earlier, possibly losing all of their social support.

In other cases, identifying as nonbinary seems to be due to social accolades that come from trying on different pronouns. For most young people, their peers are extremely supportive of any questioning of their gender. Many nonbinary people first learn about the terminology online and go through a process of changing their pronouns multiple times.

There is a larger conversation that we aren't having, and identifying as nonbinary is a way to avoid having it. Discomfort with your gender and who you are can be caused by all kinds of things that may have nothing to do with identity. If we could have this conversation openly and these individuals still chose to identify this way, it would be less concerning. But no one is saying to them, if you feel you don't fit in, it's okay to be a different kind of woman or a different kind of man.

One final video consisted of a person who identified as AMAB describing the difficulty in coming out as nonbinary and finding clothes that adequately conveyed their nonbinary status. They lamented about not wanting to wear dresses every day and voiced frustration that putting on any other garment of clothing resulted in people assuming their gender identity was male.

Another person described feeling dysphoria whenever other people identified them as male *or* female. The difficulty, however, in being recognized as "nonbinary" or expecting that people use "they/them" pronouns intuitively upon meeting without explicitly stating their preference, is that a person's sex is one of the things human beings recognize about one another.[6] Research has shown that, within a fraction of a second of looking at someone's face, the brain automatically registers information pertaining to their sex.[7] The idea that we can't—or shouldn't be able to—tell a person's gender based on their presentation or gender expression is unrealistic.

Some nonbinary people are "pronoun indifferent," which means they are comfortable being referred to by any and all pronouns, but will feel "dysphoric" if people use the one referring to their birth sex too often, or if they don't mix up the genders frequently enough. Others have opined that being misgendered as "he" or "she" instead of being called "zie" or "them" is just as hurtful as when people with gender dysphoria are identified as their birth sex. This intense concern over how they are perceived by others sounds more like social anxiety taking the form of a preoccupation with gender, which is another common thread that isn't being discussed.

Beneath all this lies a growing authoritarianism about gender and gender roles, and the assumption that accepting the gender binary requires rigid adherence to them. There's no reason you can't be a feminine man or a masculine woman, but the nonbinary label upholds the idea there's only one way to be a man or a woman. The way I see it, gender roles are only detrimental if they are limiting you, but there is no way to overcome these societal limitations if men and women keep abandoning ship and identifying as another gender.

Gender norms aren't inherently harmful. I have yet to see "femininity" or "masculinity" listed as mental disorders in any psychiatric manual. Nevertheless, femininity in women is derided and masculinity in men is mocked, but if a person takes on the label of being gender nonbinary, suddenly both are acceptable, no matter which direction they go.

There is nothing maladaptive about being gender-typical, despite all the hype around so-called toxic masculinity. Masculinity in itself does not lead to mental health issues. A meta-analysis conducted in 2017 of seventy-four studies demonstrated that conforming to masculine norms has a greater influence on whether someone seeks help when they are experiencing psychological issues, as opposed to specifically influencing their mental health.[8] It's not correct to assume that masculinity in itself increases the incidence of men experiencing mental health problems. Regarding cases in which men act violently, this is more likely explained by antisociality,* as opposed to being a masculine man.

What people now call gender could more aptly be considered personality or mood. Many seem to also be mistaking gender identity with gender expression, which is much more varied.

It reminds me of quizzes in women's magazines to find out the summer pop song that best describes your personality type, or the best on-trend beauty products for your lifestyle. Back then, you'd tally up the letters you'd circled, have a good laugh at whatever mascara or song apparently embodied you best, and carry on with your day.

But now, instead, it's all about gender, and the quizzes are not at the back of lifestyle magazines, but legitimate questionnaires as part of mandatory HR diversity training for university faculty and staff, encouraging self-contemplation about one's gender. Questions about one's gender identity and expression, offering "other" as a response along with the usual options of male and female, have become common practice. University employees are also required to select answers that state gender is fluid and that the binary model is outdated, in order to pass the module.

I liken it to any other form of self-expression. For example, I love fashion and use clothing as a way of expressing how I feel. I generally wear all black for work as a throwback to when I used to be punk, but

*antisociality: a disregard for the well-being of others.

on my days off, I look like I belong in an aerobics video, wearing lots of crop tops, high ponytails, and neon. My decisions about what I'm going to wear that day are based on a variety of factors, like how I'm feeling in the morning as I stare into the disarray that is my closet, how much sleep I had the night before, and what my schedule consists of. Sometimes other people think my clothing choices are weird or even ugly. Should public policy and language be enforced to ensure everyone respects my choices?

There is a big difference between being accepting of gender nonconformity, as I think we all should be, and enshrining into law the acceptance of labels and identities that have no evidence for their existence and are purely subjective. Those who exist along the gender spectrum report their gender changing frequently and unpredictably, without any warning or explanation, and more important, usually without their understanding.

I see nothing wrong with allowing young people to experiment with their identity by adopting these labels. It is concerning, however, when that experimentation is written into law and begins informing medical practice. For example, a 2018 study in the *New England Journal of Medicine*, one of the most prestigious medical journals, described the need for medical professionals to become aware of the unique needs of nonbinary patients.[9] A person shouldn't be considered a bigot for questioning whether nonbinary genders are real.

As the future takes another step forward on its way to becoming devoid of gender, the number of individuals taking on the nonbinary label remains a symptom of a larger issue, one that won't be solved by removing the binary. There is a worrisome amount of overlap between the nonbinary phenomenon and rapid-onset gender dysphoria (see Chapter 5).

One side of the culture war has been using pejoratives like "trans-trenders," in an attempt to move the discussion forward; the other side is pushing for over-acceptance and an attitude of "anything goes." Neither is helpful, but only one side is being held accountable. In ten years or fewer, I'm hoping the damage will be minimal as these individuals grow into their skin and inevitably return to embracing the binary.

THE FRONTIER

MYTH #4

SEXUAL ORIENTATION AND GENDER IDENTITY ARE UNRELATED

Nowhere am I more at home than at a gay club. Some of my fondest memories from my youth involve chaotic nights spent with my friends, belting out pop anthems on overcrowded dance floors, starting petty fights with their exes, and holding countless outfits in place with double-sided tape. The gay community has always supported strong women, and spending my formative years surrounded by gay men made me the woman I am today. It was through them that I learned I could be independent and outspoken, doing it all in six-inch heels.

Unlike my straight male friends, who would unceremoniously disappear the minute they got a girlfriend, my gay boys had my back, no matter what. They gave me frank advice about sex, love, and going after what I wanted in life. Being immersed in gay culture taught me a lot about thinking differently about gender and subverting expected norms. At a time when it was socially unacceptable for a woman to be assertive and stand up for what she believed in, the gay community accepted and embraced those of us who were different.

Back then, I would be the only straight woman in a sea of shirtless, dancing gay men. As my friends and I grew up and our nights out tailed off, I made a promise that I would do everything I could to protect the community. The first op-ed I ever wrote, about gender transitioning in children, was driven by this promise.

You may be wondering what my friendship with gay men has to do with being transgender—I did, after all, point out the difference between sexual orientation and gender identity a few chapters ago. Indeed, gender identity is not the same thing as sexuality, but the two are linked, and just to add further confusion to the discussion, sexual orientation is biological, but the way it expresses itself in relation to gender is influenced by social factors.

Let's start with the "sexual orientation is biological" part. Some of you may already be mad, wondering, "What proof does she have that sexual orientation is biological?" As one reader indignantly typed at me, "I don't believe Ms. Soh has discovered the gay gene yet."

Before I answer these questions, I need to lay out some background. Gay activists have historically fought for the idea that gay people were "born this way," and that being gay was not a choice, but something they realized about themselves from a young age that could not be changed.

On the opposite side of this debate has historically been the religious right, who argued that being gay is a lifestyle choice and due to someone's upbringing. This was the justification for conversion therapy, also known as reparative therapy, which aimed to turn gay people straight and has since been discredited because it is harmful and ineffective. (This is different, however, from therapeutic approaches that seek to understand a child's gender dysphoria instead of automatically facilitating early transitioning, which have inappropriately been labeled as conversion therapy and subsequently banned in twenty states, including Utah and Virginia; see Chapter 5.)

The idea that being gay is not a choice seemed to have (rightfully) won the argument, helping gay people achieve marriage equality in the United States in 2015 and the right to serve openly in the military four years prior to that.

But a new, bizarre strain of thinking has sprung forth. Due to the trendiness in conceptualizing gender and sex as spectra, the concept of "sexual fluidity" claims that anyone can be gay, and that human sexuality is in actuality free-floating and whatever you want it to be. Viewing sexual orientation as innate is considered an obsolete and oppressive mind-set that limits our self-expression and freedom.

This also fits into the debate around whether gender is a social construct. I've always said that if gender is a social construct, then sexual orientation is a *choice*, the latter of which once horrified most progressives. It seems the times have changed, because the wokest among us would now simply respond to such a challenge with "And?"

As for sexual fluidity, if sexual orientation is flexible and changeable rather than rigid and fixed, doesn't this also mean that anyone can be made to be straight? How is it not an argument in support of conversion therapy?

This ideological 180-degree change is presumably an attempt to widen social acceptance for people who are gay and bisexual, which I am all for. But the implications of doing so will only succeed at unraveling the very rights that gay activists struggled to obtain not too long ago.

Seeing as how terms are constantly being redefined in this day and age, it's probably useful to explain what I mean by "sexual orientation." Sexual orientation refers to sexual attraction to women, men, or both sexes. (Pansexuality refers to sexual attraction to both sexes, including transgender people.) Although it has been conceptualized along a variety of different domains, including attraction, behavior, and self-identification,[1] for all intents and purposes, attraction is the most accurate representation of a person's true orientation because identity and behavior can be context-dependent.

For example, someone may be attracted to the same sex, and therefore be gay, but still be engaging in sex with the opposite sex because they are not yet out of the closet. In other cases, an individual may have sex with someone as a proxy for their preferred choice of partner. The most obvious illustration of this is in prison, where there is lots of same-sex sex going on, but the inmates taking part in it aren't necessarily gay.

The Neuroscience of Being Gay

As for what the science says, sexual orientation is inborn and unchangeable. To this day, some people still believe being gay is a choice, but growing up, all of my friends argued it was innate. They had been sexually attracted to men for as long as they could remember and couldn't fathom why anyone would *choose* to be a member of a persecuted group.

The work of Ray Blanchard, a world-renowned sexologist and professor of psychiatry at the University of Toronto, speaks to the biology of being gay. His theory of the fraternal birth order effect offered an explanation, establishing that gay men were more likely than heterosexual men or lesbian women to have a greater number of older brothers. An estimated 15 to 29 percent of gay men owe their sexual orientation to this effect.[2]

According to Blanchard's research, the fraternal birth order stems from the prenatal environment. When a woman becomes pregnant with a male fetus, her body interprets it as a foreign substance due to antigens produced by the Y-chromosome. This sets off an immune response in her body, with antibodies rendering the masculinizing process inoperative, a response that strengthens with each subsequent male child. Blanchard believed this maternal immune response increased the chances that younger-born sons would be gay.

Over the years, study after study has confirmed the link between the fraternal birth order effect and male sexual orientation. In one study, Blanchard and a team of scientists, led by Anthony Bogaert at Brock University in Canada, demonstrated that the effect is indeed immunological.

In the study, mothers who had gay sons—particularly those with gay sons with older brothers—had higher levels of antibodies against NLGN4Y (a protein involved in brain development in males) than did mothers who had heterosexual sons or no male offspring. This led to differences in the way the baby's brain is masculinized in the womb.

This immune response is likely only one factor among many that affect male sexual orientation, since not every gay man has older brothers, and not every man who has many older brothers is gay. But these findings fit in with the larger scientific consensus, including neuroscientific studies, showing that sexual orientation is based in biology. (After following Blanchard's work for many years, I feel fortunate to now consider him both a mentor and a friend.)

As for the research showing that gay brains are different from straight ones, we can start by looking at the formative work of Simon LeVay. LeVay was one of the very first sexual neuroscientists, whose research in the early 1990s showed differences in brain structure associated with being gay.

In a seminal study published in the journal *Science*, LeVay, who was a researcher at the Salk Institute for Biological Studies, found that a tiny part of the hypothalamus (called the third interstitial nucleus of the anterior hypothalamus, thankfully abbreviated to "INAH-3") was more than twice as large in straight men as in gay men; in gay men, it was closer in size to that of straight women.[3] The hypothalamus is responsible for the four F's: feeding, fighting, fleeing, and fornicating. (I like to keep things PG in case there are children reading.)

LeVay's finding suggested that gay men's brains were partially feminized, and also, that the foundation of sexual orientation is established by biology. The study blew up, igniting a media firestorm around the world.

I met LeVay while attending a sex research conference several years ago. I still remember the moment vividly. It was the last day of the three-day lineup, at lunchtime, and I had been searching for the right moment to interject myself into his line of sight.

As any graduate student soon realizes, there is never a good time or a graceful way to introduce yourself to a big-name researcher who has no idea who you are and likely no interest in that changing. Any form of communication feels as though you are accosting them without warning or descending upon them like a vulture. I made several quick mental calculations as LeVay picked a few items from the lunch buffet, figuring out how much time I had before he returned to his seat, and decided to trap him in a corner as he perused a selection of soft drinks.

LeVay was seventy-one years old at the time and a legendary sexologist. He was also the author of more books than I can count, including the popular *Gay, Straight, and the Reason Why*, a book my gay friends constantly gushed about. Another reason I respected LeVay was because he had come out as a gay man at a young age, and remained out throughout his entire career as an academic.

Upon introducing myself to him, much to my surprise, he was warm and funny, and didn't seem to mind my interruption. He invited me to sit with him and we spent the lunch hour talking about his experiences growing up in England, his love of bicycling, and his partner of twenty-five years, who was a drummer in a prominent heavy metal band.

After the conference ended and I returned to Toronto—and LeVay to Los Angeles—I called him, curious to know more about his life. When I asked him about the public response to his study when it first came out, he told me he was called "reductionist" and a "phrenologist." It seems those critical of biological explanations are not particularly creative, as those are the same insults I've faced, almost thirty years later, for defending similar research.

Perhaps understandably, LeVay's work provoked fears that this ability to tell gay and straight people apart could one day be used to selectively abort gay fetuses, a worry that continues to underlie much of the criticism of research on sexual orientation today. When I asked LeVay about this, he agreed that the concerns are legitimate.

"I live in an echo chamber in West Hollywood, where everyone is gay. You are gay unless proven otherwise," he joked, before turning serious. "I actually am somewhat concerned about this issue. I do think there are things to worry about here. Maybe some New York couple might decide they'd love to have a gay kid after all but that doesn't mean that a highly religious couple somewhere are going to have the same view.

"The possibilities of misuse are actually there. It's real and I don't deny that."

When I asked him what the solution should be, he said sex researchers should continue to do more science. "I think the best way to deal with this is not just to ban the research or ban abortion or ban prenatal testing. The reason I think that is because I think parents, particularly mothers, are the best people to judge their reproductive decisions. I don't think the state should pass laws saying what women should or should not do regarding having a pregnancy or terminating a pregnancy."

He added, "The way to deal with that is to try and create a world where people won't want to do that, [to] create a world where people will feel blessed to have a gay kid. And I'm sure that a lot of people will say that's 'pie in the sky' and that'll never happen, but I see it happening. When I think about when I was a gay teenager and I compare it with now, it's unbelievable. [Society] changes. Totally unbelievable. No one could have predicted it."

I asked LeVay what it was like to be out as a gay man publicly at a time when society was still very homophobic. He spoke of how, throughout his time as a student and then as a scientist at Harvard Medical School, he had been out to his colleagues and bosses and, within academia, no one seemed to have a problem with it. "When my research came out in 1991, I never even thought, for one moment, that anyone would ask me about my personal life. I never thought journalists would be interested in it. I had no idea that that study would attract any attention. And it did attract a lot of attention.

"I was talking with some TV program in New York, one of those morning shows. I was in my lab and talking into the lens of this camera and this guy was asking about the facts and he said something like, 'I understand your being gay has affected why you did your research.' I was kind of flabbergasted—'My God, how did that guy know I'm gay, and why is he asking this?' And I said, 'Well, I am. I'm sure that's why I'm interested in the topic, part of the reason.' So that was 'coming out' for me, in terms of in the media. And I never looked back, really. There was no reason to."

He spoke about the difficulties in acquiring information about sexual orientation as it related to the brain because at death, a person's sexual orientation was rarely documented on their medical chart. The only reason he knew which individuals were gay was because they were men who had died of AIDS. It was even more difficult to ascertain women's sexual orientation.

"My original plan was to look at all four groups: gay men, straight men, lesbians, heterosexual women. I looked at any number of medical charts of women who had died and come to autopsy. I never found any reference to their sexual orientation, so I couldn't do that part of the study. I think that is probably still true today. But now, we have scanning technologies that have opened up all kinds of possibilities that weren't there before."

LeVay's partner around this time, an emergency room physician, passed away in 1990 after a four-year battle with AIDS. LeVay told me about how emotionally difficult it had been to be collecting tissue samples for the study while still in the process of grieving his partner's death.

"Most of the gay people that I studied had died of AIDS, so it was pretty painful to think of all these people, lying on flats, who had gone through the same thing that my dead partner had gone through just a year earlier."

Critics have questioned whether LeVay's findings were confounded with the fact that the gay individuals in the study had all died of AIDS. This is unlikely, however, considering that LeVay included six heterosexual men who had also died of AIDS in the sample. These straight individuals showed no difference in the size of the INAH-3 compared with the other straight men who had died of causes unrelated to AIDS.

Wanting to test the veracity of this criticism, LeVay returned to the lab after publishing the paper to examine the brain of a gay man who was HIV-negative and had died of lung cancer. That brain similarly matched the pattern found in the other gay men.

That conversation with LeVay stayed with me long after we spoke, because it exemplified what it means to be a sex researcher—the commitment to science and disproving your own bias, a curiosity to understand the world, and the desire to pursue work that was personally meaningful regardless of what society thought of it. Almost three decades later, his work still has tremendous relevance, not only regarding the politicization of science, but also the conversation relating to gender identity and being transgender.

Even if you aren't persuaded by LeVay's research, there is an entire body of sexual orientation studies showing associated differences in brain *function* waiting for you. For example, functional MRI has shown differences in patterns of brain activation between gay and straight men when viewing gay and straight pornography. The network of brain regions that are activated when gay men look at gay porn is the same network activated when straight men look at straight porn, in regions like the thalamus, which regulates sexual behavior, and the anterior cingulate, which regulates emotion.

Previous fMRI studies have also demonstrated a similar trend for gay women.[4] Regarding bisexual people, a recent study published in *Nature's Scientific Reports* compared the activation patterns of gay, straight, and bisexual men.[5] Unlike gay and straight men, who displayed activation patterns indicative of sexual arousal when looking at images and videos of men or women, respectively, bisexual men showed similar patterns when looking at both sexes.

Research has shown that people are more likely to be in favor of equal rights for nonheterosexual people after being presented with evidence that being gay was not a choice, but innate, a finding that has been stable since the 1970s.[6] Individuals who believed sexual orientation was inborn were less likely to believe gay people should have to change their sexual orientation, and were more likely to allow them to be around children without fears that they would "turn" kids gay. It also offered evidence that being gay wasn't the result of experiencing abuse. We see the same arguments being made in the current discourse on transgender rights, but these two contexts are not the same (see Chapter 5).

I don't believe being gay—or transgender—should have to be biological in order to be acceptable. People should be entitled to equal rights, regardless of whether sexual orientation or gender identity is hardwired or a choice. That gender is in fact biological won't stop individuals from identifying as the opposite sex for nefarious purposes (see Chapter 6). But in the battle against biology, one of the unexpected side effects will be the undermining of a community that activists claim to be helping.

These claims have implications that stretch beyond whom a person should be allowed to sleep with. If sexual orientation is fluid, flexible, and choosable, then so is gender identity. This works against the argument that adults should be supported in transitioning. It also counters the belief that those who have transitioned are no different from the sex they identify as.

Childhood Gender Nonconformity

As for how sexual orientation is expressed via gender identity, here is the missing piece: research has shown that childhood gender nonconformity (CGN) is one of the strongest predictors of being gay in adulthood. Gay adults have higher rates of recollecting CGN when asked about their childhood.[7] Basically, gay women recall being masculine girls, and gay men recall being feminine boys.

Despite what advocates for social justice might say, sexual orientation is linked not only with gender identity, but also with gender expression. One study showed that roughly 75 percent of boys demonstrating CGN will grow up to be gay or bisexual.[8] The reasons for this return to the prenatal environment and the extent to which hormones masculinized the developing brain.

Gender identity is flexible in prepubescent children and grows more stable as a person reaches puberty and enters adulthood. For a young child who is gender-atypical, this may be indicative of feeling as though they are more like the opposite sex upon becoming sexually mature, *or* it may be predicative of being gay.

What is the determining factor when a child reaches these crossroads? I'll get to that in a second.

Gender nonconformity and the tendency for gay people to appear more like the opposite sex—for example, for lesbian woman to appear more masculine, on average, than straight women, and for gay men to appear more feminine, on average, than straight men—may be a stereotype, but I believe we can acknowledge differences between gay and straight people without being homophobic.

The roots of the idea that gay people appear gender-nonconforming go back as far as the 1860s. Back then, Karl Heinrich Ulrichs, a German lawyer who was also publicly out as gay, devised a theory that lesbian women were male souls trapped in women's bodies and that gay men were female souls trapped in men's bodies.

Although much has changed since the nineteenth century, there remains a kernel of truth to Ulrich's broader idea. His description could be confused with our contemporary understanding of what it means to be transgender, and this is no coincidence. As we will see, neuroscientific studies have shown that the brains of gay women and men are, respectively, partially masculinized and feminized, and this has implications for modern-day brain imaging studies involving transgender people.

Greater exposure to prenatal testosterone is associated with male-typical interests and behaviors and sexual attraction to women, regardless of whether the individual is male or female. Boys, for instance, are typically exposed to higher levels of testosterone in the womb, and tend to gravitate toward mechanical toys, like trucks, and being sexually attracted to women upon reaching puberty. A boy who is exposed to lower levels of testosterone is more likely to be *female*-typical when he is born, gravitating toward toys and activities that girls prefer, like dolls and playing house, since girls are also generally exposed to lower levels of testosterone. He will also be sexually attracted to men in adulthood.

Across a variety of animal models (including rats, mice, and ferrets, since it's not ethical for us to use human fetuses as test subjects), changing the amount of testosterone that an animal is exposed to changes whether they are sexually interested in same-sex or opposite-sex mates.[9]

Another piece of the puzzle becomes evident through studies involving girls with congenital adrenal hyperplasia (CAH). As discussed in Chapter 2, these girls are exposed to higher than usual levels of testosterone in the womb. Roughly 3 percent of girls with CAH will grow up to identify as male,[10] and of those who don't, a large proportion will identify as lesbian.[11] The opposite is also true; children whose mothers took anti-androgenic medication (which lowers their testosterone levels) during pregnancy tend to prefer female-typical toys.

Rough-and-tumble play, which is behavior typically seen in little boys, is less common in gay men and more common in lesbian women, due to lesser and greater masculinization of the brain, respectively. Whenever I explain these differences to an audience, I'm reminded of one of my former colleagues in sex research, who is gay. He used to tell me stories about how, from a young age, he knew that he and his sister, who is also gay, were different from other children. He was not like other little boys, who were rambunctious and aggressive in their play fighting, and she was always much more like them.

The Neuroscience of Being Transgender

I'm often asked whether the brain of someone who is transgender is different from the brain of someone who isn't, and if so, do I think we will one day develop a "brain scan test" to tell one from the other?

Regarding the first part of the question, the answer is yes and no. Research suggests, in addition to potential differences in the brain, there are other factors at play influencing a person's feelings of gender dysphoria and their decision to transition, including the kind of sexual partners one seeks to attract.

I'll start with the part about brain differences. Much of the early neuroscientific work involved postmortem studies. The tricky thing is, most of the individuals in these studies had been taking cross-sex hormones, which can alter the tissues in the brain and make it difficult for scientists to know whether the brain differences they were seeing were due to a biological condition or the hormones that an individual was taking.

Thankfully, advances in brain-imaging technology have offered new insights into understanding gender dysphoria. For example, one study from 2018 gained widespread headlines for suggesting that the brains of gender-dysphoric children are more similar to the sex they identify as, as opposed to their sex at birth, potentially offering support for the early transitioning approach.[12]

It's important to mention, however, that participants' sexual orientation was not reported, so it becomes difficult to extrapolate what these findings really mean. The brain differences the researchers found *could* be due to the children being gender dysphoric *or* due to being gay. Since gender nonconformity and gender dysphoria have a tendency to present in children who later grow up to be gay, it's not a far cry to presume the children in the study might be attracted to individuals who share their sex at birth, and therefore, the brain differences pertained to sexual orientation, *not* gender identity.

In another study, from 2015, researchers measured cortical surface area, comparing three groups of children aged twelve to seventeen years old: eleven hormonally untreated, gender-dysphoric children who were born female, with five nontransgender girls and five nontransgender boys. The researchers found that the brains of the gender-dysphoric girls were shifted in a male-typical direction; they were more male-typical than girls without gender dysphoria, but less male-typical than boys.[13]

I mentioned earlier that gender identity is flexible in prepubescent children and that it grows more fixed with age, and with puberty comes a surge of hormones that promotes sexual dimorphism in the brain. Participants in the aforementioned study have undergone puberty, but it is possible they aren't out as gay yet. It's also important to keep in mind that small sample sizes were used in this particular study. In my last conversation with the lead researcher on that team, I was told they are in the process of collecting more brain data.

Similar findings have been shown in adults; transgender women and transgender men have typical gray matter volumes (or cell bodies), but differences in white matter (connective tissue). Researchers in Madrid conducted diffusion tensor imaging studies, looking at white matter in the brain, comparing transgender research participants with male and female control (that is, nontransgender) participants. Importantly, the trans people who took part in these studies were not taking hormones at the time.

The brain structure of transgender individuals appears to be shifted in the direction of the sex they identify as, as opposed to resembling the brains of those who share their birth sex. In one study of 18 hormonally untreated trans women, 19 female controls, and 19 male controls, researchers scanned the participants' brains and compared them across a measurement called fractional anisotropy, which is a fancy way of saying how easily the water molecules move through the tissue. Men tend to have higher fractional anisotropy than women. Transgender women demonstrated a white matter trend *in between* women and men, with greater fractional anisotropy than women, but less than men, suggesting that their white matter tracts were only partially masculinized during development.[14]

A similar trend was seen in transgender men. In a study of 18 hormonally untreated trans men, 24 male control participants, and 19 female controls, trans men showed a brain connectivity pattern closer to people who shared their gender identity (that is, men) than those who shared their birth sex (that is, women).[15]

Both of these studies' results, however, were also confounded with sexual orientation, as all of the transgender participants were gay (attracted to people who shared their birth sex), while all of the controls were straight (attracted to the opposite sex). It remains unclear whether the brain differences found were a reflection of gender identity (being transgender) *or* sexual orientation (being gay).

It's important to clarify, this is not to say that all transgender people are really just confused gay people. Research has shown that transitioning can indeed be beneficial for some people, helping them to live much happier and healthier lives. But if we want to truly understand what it means to be transgender, including whether this is something hardwired in the brain, we have to look at the question from all angles, including those that make us uncomfortable because of their potential for misuse.

Regarding the feasibility of using brain scans to determine whether someone is "really trans," right now, identifying as transgender is based purely on that—self-identification. Similar to developing a test to determine whether a fetus is gay, a brain test to determine whether someone is gender dysphoric poses ethical risks. Will it be used to "out" people? If someone says they experience gender dysphoria but this doesn't manifest as a particular result on the test, will they be viewed as malingering or experiencing symptoms that are less severe, and thus, face additional obstacles if they pursue transitioning?

Transitioning can include medical interventions like hormonal blockers and cross-sex hormones. Additional procedures for individuals born female include chest binding or having a double mastectomy, in some cases in the United States, as young as age twelve. For those born male, surgical options include getting breast implants, facial feminization surgery (which can involve cheek and lip augmentation and shaving the brow bone, nose, and jaw), tracheal shaving, and voice feminization surgery, which involves shortening the vocal cords.

Bottom surgery, also known as gender reconstruction surgery (previously known as "sex reassignment surgery," "gender reassignment surgery," "gender confirmation surgery," and "genital affirmation surgery") involves the creation of either a neovagina or a neophallus. Vaginoplasty consists of inverting the penis or using the lining of the colon to create a vaginal canal. Phalloplasty involves creating a penis from a large section of tissue taken from another part of the body.

These interventions can indeed be lifesaving for some. In order to increase the likelihood that transitioning will have positive outcomes, research in this area must be rigorous and nonpartisan, because undergoing these interventions isn't without risks. Chest binding can lead to back pain, rib and organ damage, and difficulties breathing. Vaginoplasty requires dilating, which can be very painful, to ensure the neovagina doesn't close up. Surgeries constructing a penis involve more potential complications and will leave visible scarring. Hormonal treatments can result in permanent side effects (see Chapter 5).

Some have questioned the utility of transitioning from the perspective that clinicians should work at healing the mind instead of the body. I often hear the comparison of gender dysphoria to anorexia—and transitioning to liposuction—the implication being that you wouldn't indulge someone whose beliefs about their weight were not reflective of physical reality. Instead, someone suffering from an eating disorder would be encouraged to undergo therapy in order to change their beliefs about their body.

The same has been said of body-dysmorphic disorder (a mental health disorder that revolves around a preoccupation with perceived flaws in one's physical appearance that are imperceptible to another person) and cosmetic surgery, or individuals who identify as transracial (which means they identify as a race that is not of their racial background) and altering one's skin tone.

Indeed, for some struggling with gender dysphoria, their issues with their gender would more appropriately be considered a symptom of another mental health condition. A study from the *American Journal of Psychiatry* showed that 61 percent of patients presenting with gender dysphoria have another psychiatric disorder. In 75 percent of this 61 percent of patients, gender dysphoria was a symptom of another mental illness, such as a personality, mood, or psychotic disorder.[16] A common example would be borderline personality disorder, which includes core symptoms like having a rapidly shifting, unstable sense of self, suicidal ideation, and self-harm. Others can include body-dysmorphic disorder, anxiety, autism, or an eating disorder.

This of course does not mean that for *everyone* experiencing gender dysphoria, it is a secondary symptom of another issue. But those who are against transitioning as a solution will point to these facts as evidence that transitioning is unhelpful. The misdiagnosis of another mental health condition as gender dysphoria is a problem that will only be compounded by the lack of adequate diagnostic assessments being done currently, due to trepidation on the part of clinicians that failing to affirm their patients will lead to them being fired.

For individuals whose gender dysphoria does not stem from these other conditions, these comparisons are not apt. Gender identity is an inherent part of who we are, since our experience of gender is dictated by biological influences exerted before we were born. Although race similarly has biological correlates, associated with things like different medical illness predisposition, the same can't be said for someone who is transracial, because our experience of race is not biologically determined. Progressives will, however, laugh at someone who believes they are transracial, like Rachel Dolezal,* when hypothetically speaking, one could expect transracialism to be more common than gender dysphoria because how we experience race is more subjective than how we experience gender. Nevertheless, it bears repeating, gender dysphoria is a real condition whereas transracialism is not.

A person may have a biological predisposition for anorexia or body-dysmorphic disorder or any mental health condition, but it is understood that supporting their desires only perpetuates, and does not heal, their distress. In the case of someone with anorexia, encouraging them to lose extreme amounts of weight is harmful because it can lead to death.

Some may consider transitioning to be "mutilating" healthy flesh, but for an adult who is gender dysphoric, as risky as some of the related medical interventions may be, transitioning can help to alleviate their discomfort. One meta-analysis of twenty-eight studies showed that transitioning can indeed be beneficial for some adults.[17]

Even if scientific research showed no brain correlates associated with gender identity, adults should be free to do what they want with their bodies. Some are able to resolve their gender dysphoria without making

*Rachel Dolezal: the former branch president of the National Association for the Advancement of Colored People, who resigned when it was revealed she was a white woman who identified as black.

any physical changes, but I don't believe this should be enforced as the rule for everyone.

Another argument consists of pointing to a Swedish study from 2011 that has been misquoted as saying that transgender people fare worse after transitioning and that gender reassignment surgery increases suicide. In the paper, however, the study authors argued that the outcomes for these individuals might have been worse had they not undergone gender reassignment surgery.[18]

Returning to whether gender dysphoria should be considered a mental illness, conceptualizing it this way can—and indeed has—given ammunition to those who seek to dismiss transgender people's concerns and rights with claims that they are mentally unwell, don't know what they are talking about, and shouldn't be listened to. But reclassifying gender dysphoria does nothing to combat ignorance or to increase understanding. We should be talking about how someone with a mental disorder deserves to be treated with love, respect, and acceptance instead of pretending that what they are struggling with has no bearing on the mind at all.

Calling gender dysphoria something other than a mental disorder also makes it more difficult, in the long run, for those who are suffering to get the help and support they need. Some members of the trans community, including Buck Angel, are concerned about the change. Angel identifies as a transsexual man and is an LGBT+ icon. He is a sex educator and creator of sexual and health wellness products, including the world's first sex toys for trans men. He said the mental health component of transgender care is critical, and removing gender dysphoria from being classified as a mental disorder means an individual will have to go out of their way to access counseling as part of their treatment.

"Imagine you are a car and your brain is your engine. . . . Mental health, to me, is basically like a mechanic, and they're fixing your engine to help you go and get your car painted, get your new tires put on. And so to me, when we remove mental health care, we're removing the mechanic. We're removing the mechanics to your whole being. If your brain isn't functioning properly, I don't care what you do to your body. I don't care how many surgeries you have. I don't care how many hormones you take. Now we aren't talking about mental health care as part of the equation of being trans."

I believe scientists need to be vigilant that technological advances are not used to support discrimination against the transgender community. But ultimately, we don't need to redefine gender dysphoria or rely on research findings—brain-based or otherwise—to legitimize what it means to be transgender or to afford trans people equal rights.

Autogynephilia

Now, we'll get to the ways in which sexual orientation influences the expression of gender identity, and how one's preferences in romantic and sexual partners play a role. I've been talking a lot about transgender people who are attracted to people who share their birth sex. Since not all trans people are gay, you're probably wondering, "What about trans people who are sexually attracted to the opposite sex?"

The answer to that question is *autogynephilia*, the number one reason why you will rarely, if ever, see sex researchers—or any expert, really—criticizing the transgender community.

You're probably thinking, "Auto-gyne-*what*?" and "Why is it so scary?"

Before I get to the details of exactly why everyone is so freaked out by autogynephilia, let me explain what it means.

In the debate around trans activism, autogynephilia has become a code word indicating you know what's really going on. Colleagues I've met through my work as a journalist will often whisper it to me, hesitantly.

"I've heard about . . . autogynephilia," they'll say.

"It sounds like the name of an alien," a TV producer once told me.

Ray Blanchard, the scientist I mentioned earlier who found that gay men tend to have a greater number of older brothers, originally coined the term. In the late 1980s, he would see patients with gender dysphoria in his work as a psychologist, and noticed that those who were born male fell into one of two categories. In an attempt to understand this, and particularly the phenomenon he was witnessing of individuals disclosing that they were sexually turned on by the idea of becoming a woman, he devised a typology with the goal of informing other clinicians in the field.[19] A name for this phenomenon didn't yet exist, so he devised the Greek word *autogynephilia,* which translates to "love of oneself as a woman."

Before we go any further, I'll define a few terms that are often mistaken for one another. A *cross-dresser* is a straight man who wears women's clothes, underwear, and makeup because he finds it sexually arousing. These individuals were once called "transvestites," but cross-dresser is considered a more respectful term.

Transvestic disorder, which is included in the *Diagnostic and Statistical Manual of Mental Disorders,* known colloquially as psychiatry's Bible, is a paraphilia (an unusual sexual preference) revolving around the desire to wear clothing associated with the opposite sex. Not everyone who enjoys cross-dressing, however, has a disorder. The key difference between your garden-variety cross-dressing or transvestic fetishism and having a disorder, per se, is whether these activities and urges are excessive or troublesome, leading a person to experience distress or problems in their day-to-day life. It's indeed possible to indulge in cross-dressing without it being pathological or causing harm to oneself or other people.

A *drag queen*, on the other hand, is a gay man who dresses up in women's clothing to emulate extreme forms of femininity. Some transgender women may choose to perform in drag, but by and large, most trans women are not drag queens. Performing in drag is not typically motivated by erotic desires, although the performance itself may be sexual.

Blanchard's male-to-female transgender typology consisted of two subtypes: the androphilic (which means male-attracted) subtype, and the autogynephilic (or female-attracted) subtype. The androphilic subtype is also known as the gay subtype because these individuals are attracted to people who share their birth sex. (As a refresher, in sexology, whether a trans person is considered gay or straight is based on whom they are attracted to in the context of their birth sex.) The autogynephilic subtype is known as the heterosexual or bisexual subtype, since autogynephilic individuals are typically attracted to women, or both women and men.

The first subtype described individuals who showed signs of being very effeminate from a very young age, presenting with what is known as "early-onset" gender dysphoria. From the moment they were born, they would gravitate toward playing with female-typical toys like dolls, wearing dresses, and makeup, and preferring girls as friends instead of boys. They would profess to their parents that they should have been born as the opposite sex. Upon reaching sexual maturity, they were exclusively attracted to men (hence why they were considered gay, because their sexual partners shared their sex at birth).

They usually transitioned before the age of thirty, and their desire to do so was primarily motivated by the type of sexual and romantic partners they sought to attract—namely, very masculine, heterosexual men—considering that straight men are sexually interested in women.

The other subtype, known as the autogynephilic subtype, consisted of individuals who didn't experience any feelings of discomfort around being male *until* reaching puberty. Most were male-typical and masculine throughout their childhoods—in some cases, playing competitive sports and taking up computer-related hobbies[20] with their male peers. (This is not, of course, to say that some women don't also enjoy tinkering with computers.)

These individuals usually first experimented with cross-dressing at a young age. Upon reaching puberty, they were sexually attracted to women, but also experienced sexual arousal to women's clothing and the idea of *becoming* a woman. Their gender dysphoria is considered "late-onset," with a different etiology from the gay subtype, stemming from a desire to become the women they were attracted to. Most would come out as transgender in middle age, after being married with children for many years, and with successful careers in male-dominated professions like business or computer science.

Many report being female-typical in childhood because, perhaps on a subconscious level, they are aware that doing so makes them appear more like the gay subtype. They will also deny any sexual motivations for transitioning, because denying the sexual component increases the likelihood that medical professionals will give them the green light.

You may be surprised by this information, wondering why you've never heard about any of this before, considering how prominent trans voices are nowadays, and particularly those of trans women. This is because activists and allies have committed to promoting the narrative that all transgender people across the board feel the way they do because they possess an internal sense of gender that is in conflict with their anatomic sex, a feeling that has no relevance to sexuality or their sexual orientation, in an attempt to neutralize the sexual aspect of their wishes to transition. This is not only unhelpful in the long run for people who experience autogynephilia, but it is also sex-negative.

What I *don't* want is for people to take the information in this book and use it to deny transgender people their rights, legal protections, or access to transition, *or* to use it to paint transgender women, in particular, as fetishists or sexual deviants, as has been done in the past. I don't believe it's acceptable or warranted to use autogynephilia as a point of attack when debating these issues. The fact that, for some people, transitioning has relevance to sexual arousal is not a justification for judging anyone who feels this way. This is also not to say that their experience of gender dysphoria is any less real or serious than the gay subtype.

Sexual fantasies reported by people who experience autogynephilia generally fall into one of five themes:[21] experiencing female physiological functions, like menstruating, lactating, and being pregnant; possessing female body parts, like breasts and a vulva; dressing in women's clothing, ranging from matronly blouses to sexy lingerie; engaging in stereotypically feminine behavior, like taking birth control and speaking with a female voice (and in some cases, riding a girl's bicycle[22]); and interacting with others as a woman—in many cases, taking on the receptive role during sexual intercourse with men. The key difference between the gay subtype and the autogynephilic subtype, with regard to having sex with men, is that the gay subtype is attracted to the male physique and men's bodies, while the autogynephilic subtype fantasizes about being penetrated as a woman.

Autogynephilic fantasies can additionally revolve around shape-shifting and body-switching, in which a person unexpectedly transforms into the opposite sex by way of sci-fi themed occurrences, like alien abductions, ingesting magical potions, and drifting into outer space. There exists an entire genre of erotic literature devoted to "gender swap"–themed story lines that feed into this particular fantasy, which describe the male lead character's instantaneous and often unexpected metamorphosis into a busty plaything.

Many feminists take issue with transgender women, and especially the autogynephilic subtype, alleging they perpetuate sexist stereotypes of what it means to be a woman—self-objectifying, airheaded, and only good for sex (see Chapter 6). Trans women are seen as embodying clichés about what women are supposed to look like, including long hair and nails, heavy makeup, and high heels, a "costume" that says little about what it means to be a woman, particularly in the eyes of feminists who have fought long and hard against it.

As you can imagine, information about autogynephilia doesn't sit well with trans activists, because it challenges the more sexually sanitized "woman trapped in a man's body" narrative. Blanchard named this socially pleasing narrative the "feminine essence theory."[23] When I was a hard-core feminist, I learned that transgender women were women trapped in men's bodies and that was that. I still remember the day, several years later, while in graduate school and sitting in on a case conference with a number of senior sexologists, when I realized that this belief had no evidence backing it.

When it comes to brain science, to date, one MRI study has looked at the neuroanatomy of trans women who are sexually attracted to women.[24] The sample included 24 male-to-female transgender individuals who were attracted to women, 24 women who were attracted to men, and 24 men who were attracted to women. The researchers found that, in sexually dimorphic parts of the brain (regions that are usually different between women and men), the brains of these trans women were no different from those of the nontransgender men, unlike what is typically seen among trans women who are sexually attracted to men, whose brains show a pattern that is in between male and female in sexually dimorphic areas.

Brain differences that *were* found between the trans women in this study and nontransgender men were *not* in brain regions that are sexually dimorphic. This suggests that the neural differences seen between nontransgender men and people who experience autogynephilia are due to something other than feeling female.

For this reason, we have to differentiate between the gay and autogynephilic subtypes when talking about the transgender brain. There are two very different, discrete phenomena at play that have implications for the best way forward for trans women, depending on which subtype they fall into.

Autogynephilia is frequently dismissed as "outdated medical research" or "junk science" that has been "disproven" and "put to rest." If you do any searching for information, it will consist mostly of blog posts by transgender women saying the whole thing was made up by a bunch of transphobic sexologists to invalidate trans people.

A common counterargument that is raised in an attempt to negate autogynephilia is the claim that heterosexual women experience autogynephilia, too. A 2009 study surveying twenty-nine female hospital employees stated that 93 percent of them reported some form of autogynephilia.[25]

The questions used to assess autogynephilia in these women, however, did not adequately capture what autogynephilia, by Blanchard's definition, actually consists of. For example, the study asked respondents whether they had ever been sexually aroused by dressing in lingerie or wearing makeup, but in the context of meeting a sex partner, as opposed to merely doing the action itself. Questions from this study also focused on whether the female participants became erotically aroused by wearing sexy clothing, whereas autogynephilia manifests as sexual arousal from even nonsexy attire, like nightgowns and old-school granny panties.

Beyond that particular study, autogynephilic individuals often self-report difficulties getting dressed in women's clothing without having an erection or ejaculating. I don't know of any nontransgender women who get turned on or have an orgasm as the result of getting dressed in the morning.

Like many truths around the transgender debate, information is swept under the rug, and so-called experts deny its veracity because they are either afraid of the activists or their livelihoods depend on them doing so. Within sexology, everyone acknowledges that autogynephilia is real, including clinicians who treat adult patients presenting with gender dysphoria, but only a handful of my colleagues have been brave enough to go on the record saying so.

What happens when you discuss autogynephilia publicly? In the case of Michael Bailey, a professor of psychology at Northwestern University, he very nearly had his professional and personal reputation ruined after writing a book in 2003 called *The Man Who Would Be Queen*. In it, Bailey discussed Blanchard's male-to-female transgender typology, including autogynephilia.

As punishment for doing so, Bailey endured a prolonged smear campaign at the hands of several prominent trans activists. Some of their tactics included having the book stripped of its award nomination from the Lambda Literary Foundation, publishing photos of Bailey's children on a website with "satire" that he had sexually abused them, and bombarding the psychology department at Northwestern University with emails stating, among other things, that he was an alcoholic.

They also filed several complaints to Northwestern's ethics board, including allegations that Bailey conducted research involving human participants without ethical oversight, and that he had sex with one of his transgender research participants. A very thorough investigation by Alice Dreger, a historian of medicine and a professor of medical humanities and bioethics at Northwestern at the time, concluded that all of these allegations against Bailey were false.[26]

Critics who support the position that autogynephilia doesn't really exist usually assume that those of us arguing otherwise are against trans women transitioning. Importantly, Bailey never said that presenting with autogynephilia should disqualify a person from consideration for transitioning; in fact, he has publicly said that he supports medical interventions for transgender adults who seek it. I am in full agreement with this.

But the entire ordeal left the field of sexology shaken—this is what happened when a scientist discussed autogynephilia, and unless you wanted the same thing to happen to you, it was best that you put your head down and not make a fuss.

It makes sense why there has been such a backlash to any mention of autogynephilia. Because autogynephilia is considered as a paraphilia, it is technically classified in the medical literature as a mental disorder. During my time as a researcher studying paraphilias, I would see for myself the extent to which people with atypical sexual interests faced judgment from their sexual partners and society, more broadly. Perhaps more pressingly, paraphilias cannot be changed, which adds to the anguish and shame many paraphilic individuals quietly experience.

I can also understand how it can come across as patronizing for scientists to say that they understand transgender people better than they know themselves, particularly if the scientist isn't transgender. It is also probably infuriating to be told that you have a "disorder" and what you are struggling with should be pathologized. In our sex-negative culture, trans women's concerns are dismissed as a sexual fetish if anything about transitioning relates to sexual desire.

I would argue that a mental condition should be considered pathological if it is causing someone harm or impairment, and indeed, these are some of the diagnostic criteria used in evaluating whether someone has a mental disorder. This is not to say that autogynephilia or the fact that it is sexually motivated means these individuals deserve to be treated differently from anyone else. Autogynephilia is only pathological if it interferes with a person's ability to live their life. If that was never the case for anyone, then why do some trans women seek medical interventions like transitioning? The problem is not in labeling something as pathological, but that psychopathology is stigmatized.

I advocate for compassion and not being judgmental. I can only imagine how distressing it must be to have the realization that this—or anything besides vanilla, heterosexual sex—is what turns you on, knowing full well how taboo it is in society. In some cases, a child who has just reached puberty will be brought by his parents to see mental health professionals after being found with his female relatives' underwear in his possession.[27]

In today's political climate and with access to Internet forums, more autogynephiles are self-diagnosing as transgender at younger ages. When we consider the number of referrals to gender clinics, the number presenting with the autogynephilic subtype has increased in the last thirty years. In 1987, roughly 60 percent consisted of the autogynephilic subtype compared with the gay subtype; in 2010, it had increased to 75 percent and is likely even higher now.[28]

It's important to note that not all autogynephiles identify as women. Some acknowledge their sexual fantasies but are content to live and identify as men, or to live in the female role in private, only part-time. This isn't to say that *all* individuals who are transgender, or even all of those who experience autogynephilia, are better off not transitioning. But it is a relevant point to consider, if we are looking for the best outcomes for those questioning their gender.

From a therapeutic perspective, my colleagues have suggested that clinicians require transgender women considering surgery to live as a woman for at least one year, or preferably two, before going ahead with it.[29] Those who have studied autogynephilia tell me that many individuals who feel this way reach out to them, thanking them for helping them understand themselves.

I want to be clear: I bring up autogynephilia because it's a critical part of the discussion on gender transitioning, and particularly the conversation about whether young children should be allowed to transition (see Chapter 5), even though the two groups are *not* the same. We need to be able to have an honest conversation if we want to effectively help people suffering from gender dysphoria.

Autohomoerotic Gender Dysphoria

As for transgender men, there is some overlap with Blanchard's male-to-female typology, in that most individuals who are born female and transition to male will experience gender dysphoria that has a childhood onset and will be gay (that is, in this case, sexually attracted to women). Less has been known about trans men because, as discussed, referrals to gender clinics have historically been mostly for trans women. Up until about ten years ago, the majority of children and adolescents seen were individuals who were born male.[30]

This sex ratio has inverted more recently, and gender referrals among youth are primarily among those born female wishing to transition to male (see Chapter 5 for a discussion of rapid-onset gender dysphoria). As greater attention and awareness is brought to issues affecting trans men, the corresponding number of studies will hopefully increase, as well.

As for transgender men who are sexually attracted to men, there does not exist a phenomenon analogous to autogynephilia. This is probably because paraphilias are primarily found in people who were born male. (The only exception to this is sexual masochism, which is more commonly seen in women.[31])

There is an emerging trend of young women who choose to transition to male because they are sexually attracted to gay men. They are heterosexual, having been born female and experiencing sexual attraction to men.

Blanchard calls this *autohomoerotic gender dysphoria*, which is sexual arousal at the thought of being a gay man. It isn't, however, the female version of autogynephilia, for several reasons. Those who are autogynephilic are turned on by picturing themselves as having the body of a woman, while individuals with autohomoerotic gender dysphoria are turned on by the thought of having *gay male sex*. With autogynephilia, becoming a lesbian woman is not the focus of transitioning, but a secondary outcome, whereas with autohomoerotic gender dysphoria, becoming a gay man *is* the focus. Sexual cross-dressing is also not present in autohomoerotic females.

The fact that we can't talk about the role of sexual orientation in the context of transitioning is doing an enormous disservice to people struggling with gender dysphoria. Nowhere is this more apparent than in the debate on childhood transitioning.

MYTH #5

CHILDREN WITH GENDER DYSPHORIA SHOULD TRANSITION

While at an event a few summers ago, I found myself in the usual predicament of figuring out what to do with the drink in my hand. There had just been a toast, photographs were taken, and now, as someone who had quit drinking and who is also very Canadian, I was trying to discreetly dispose of the plastic flute of champagne I'd been given without seeming rude.

It was dark in the room, so I had that working in my favor. I considered offering my glass to a colleague, but in my experience, this usually prompted concerned questions about whether I was pregnant and who the father was. Just as I managed to disappear to one side, having located a large trash can hiding in a lonely corner, I heard a man speak my name.

I thought he was about to ask me why I was wasting a perfectly good drink, but he instead told me that he had a transgender daughter. She had been born male, but always gravitated toward girls' things, and from the moment she could speak, proclaimed she was really a girl. She would wear a towel over her head, pretending it was a cascade of beautiful, long hair, in addition to sneaking into her mother's dresses and collection of nail polish. She hated being a boy and constantly asked why she couldn't become a girl. After much consideration, he took her to a gender specialist and she was now on puberty blockers.

I nodded as he mentioned the clinic he was taking her to. It turns out I knew of the doctor she was seeing, someone who was pro–early transitioning and also not terribly scientific, in my opinion. I couldn't find the words to tell him this. Parenting decisions are such a personal thing, and I wasn't about to criticize someone for what he thought was an appropriate choice.

I explained what my conclusions were based on, summarizing a few research studies and explaining how politicized the subject had become. In my mind, it was very likely that he did not have a transgender daughter—he had a gay son.

He said the treatment his daughter was receiving was cutting-edge, and that the likelihood for depression and suicide was high if she did not transition.

I realized we had come to an impasse. It wasn't as though he was going to reconsider her path, and why should he? I had never even met his child. Why would he trust me, a total stranger, over an entire medical team that presumably knew her well? But I could see the pain in his eyes and knew there was a part of him that felt torn.

I've had many parents approach me over the years, asking if their decision to allow their child to transition was the right one. In this chapter are all of the things I wanted to tell them.

The "Myth" of Desistence

Across all eleven long-term studies ever done on gender dysphoric children, between 60 and 90 percent desist by puberty.[1] Desistence refers to the phenomenon of gender dysphoria remitting. A child who has desisted will no longer feel dysphoric about their birth sex. These kids, who would fall into the early-onset, gay subtype of Ray Blanchard's transgender typology, are more likely to grow up to be gay in adulthood, not transgender.

Regardless of whether we look at older or newer studies, no matter how large or small the sample size, or where in the world and which research team conducted it, the data are irrefutable. But you wouldn't know it based on the number of "expert" sources claiming desistence has been disproven. Every few months, those of us defending the science must go through the process of explaining why it is legitimate for yet another time, and another time, and another time after that.

I agree that questioning scientific findings is a good thing, because it enables us to get closer to the truth. If the data showed something different, then we should be open to reevaluating our opinions. But in this case, critics are committed to finding fault with these studies because the results do not support their political end goal.

In the many years I've been following this issue, I've seen how discussions about gender dysphoria have become increasingly policed, not just by activists, but medical organizations, academia, the media—and most alarmingly—parents, too. Now anyone who offers a nuanced counterpoint or hints that maybe we should take a step back and slow down puts themselves at the mercy of a scorched earth reprisal.

What leads to desistence? Upon reaching puberty and developing romantic interest in their peers, the majority of children who once felt discomfort with their bodies grow to be comfortable in them.[2] This is because gender identity is flexible in prepubescent children and grows more stable with development and age.

The logical conclusion would then be that clinicians should advocate for a cautious approach when advising their young patients on whether or not they should transition. Instead, this body of research has been cast aside as "bad science" peddled by anti-transgender folks, an allegation stemming from understandable apprehension that desistence may be used to justify bias against transgender people.

I understand the desire to hide this information. Acknowledging that some transgender people change their minds could be misused to support claims that trans people don't really exist, that what they feel isn't real, or that efforts to support transitioning should be halted across the board for everyone. The transgender community has faced medical gatekeeping in the past, including difficulties accessing the care and support they deserve.

But overcorrecting for the past won't help gender dysphoric children today, especially for those who will end up desisting and changing their minds, as we will see when we get into the discussion of detransitioners and transition regret.

Because so much misinformation exists, and because many trans people themselves have had negative experiences with the medical community, some also likely believe that the research itself is biased. There are several main criticisms that desistence deniers usually bring up when debating this issue. I have sought out counter-information to my position in an attempt to prove myself wrong. The last thing I want to do is make life harder for the transgender community and these children. I've read all of the think pieces claiming the science is spurious, and every time, their points ring hollow.

If you've read anything about gender dysphoric kids, you've probably encountered the catchphrase "consistent, persistent, and insistent." It suggests that children who are *truly* transgender and would benefit from transitioning have an unshakable—or consistent, persistent, and insistent—sense that they are the opposite sex.

But really, aren't all children consistent, persistent, and insistent when they want something badly? Kids say all kinds of things, and yet, adults don't take them seriously about any of it *unless* it relates to their gender. We don't allow children to get tattoos or piercings, to drive, vote, drink, or buy cigarettes, because they lack the emotional maturity to make life-altering decisions.

Another criticism is that the diagnostic criteria for gender dysphoria, as defined by the American Psychiatric Association's *Diagnostic and Statistical Manual of Mental Disorders* (DSM), came about only in 2013, and children diagnosed with what was then known as "gender identity disorder" (defined in 1994) exhibited less severe symptoms than what would be required to meet a diagnosis of gender dysphoria today. The newer criteria placed a greater emphasis on a child believing they were the opposite sex in order to be diagnosed.

Because most of the studies on desistence were conducted prior to 2013, activists claim that these two diagnostic populations—children diagnosed using the previous criteria and those diagnosed with the current—are different, and desistence findings have no relevance to children who are *really* gender dysphoric.

But in many cases, children will say they are the opposite sex because they want to do the things that the opposite sex does, and this is the only language they have to communicate this. It is only around the ages of five to seven years old that children learn that gender is a fixed characteristic that doesn't change based on superficial things like appearance (for example, wearing a dress) or activity (say, wanting to have a tea party).[3] And even when you remove kids who are less severe from the study analysis, the rate of desistence is still over 80 percent.[4]

Even with more stringent criteria, it's still possible that children who say they *are* the opposite sex will end up desisting one day. Severity of gender dysphoria—presumably what is being captured by the more recent diagnostic criteria, and what activists are flagging—isn't necessarily predictive of a child's dysphoria persisting. As one study showed, even children who were severe in their gender dysphoria desisted.[5]

As well, from what I've seen, children are being given a diagnosis without a strict adherence to the diagnostic criteria. Clinicians have much leeway to use their own judgment when it comes to determining whether a child meets the cutoff for a diagnosis. In some cases, a child experiencing gender dysphoria won't even undergo a full assessment, never mind one that is thorough, before being given access to medical treatment.

This is the sentiment of a lawsuit being led by Keira Bell, age twenty-three, against a clinic in the United Kingdom. The lawsuit challenges the clinic's practice of prescribing puberty blockers and cross-sex hormones to children under the age of eighteen. As a former patient of the clinic, Bell transitioned to male, starting puberty blockers at sixteen and testosterone shortly thereafter. After undergoing a double mastectomy at twenty, Bell regretted the decision, and like many detransitioners, as we'll discuss, she has since returned to identifying as female. Describing the process, Bell said she was prescribed puberty blockers after three one-hour appointments and experienced little resistance.[6]

Yet critics will continue to say that children who desisted were never truly gender dysphoric; they were only "gender-nonconforming." This is hypocritical, though, because they would never say that of anyone else who is questioning their gender, as that would be considered invalidating someone's "lived experience." We see the same thing happening to girls with rapid-onset gender dysphoria (which we'll be getting to), detransitioners, and men who self-report being autogynephilic (see Chapter 4). "Lived experience" matters only if your story fits the narrative.

Another criticism has been study attrition—the largest longitudinal study on boys and girls with gender dysphoria, conducted in the Netherlands, lost a number of children at follow-up.[7] Since this clinic was the only one in the country providing gender services, the researchers assumed that those who didn't return for hormones or surgery had desisted. Critics argue, however, that we have no way of knowing that for certain—it is possible they pursued medical intervention elsewhere or decided to live their life as a trans person without undergoing any medical procedures.

If we are charitable and go to the numbers directly, however, even after removing the number of children lost to study attrition from analysis, the desistence rate was still more than 50 percent, which suggests it is still a high-probability outcome.

You may be wondering about the small minority of kids who don't desist. I didn't, after all, say that 100 percent of gender dysphoric children outgrow their feelings by puberty. So, what about the 10 to 40 percent of children who will persist?

Indeed, there is a small percentage of children who, upon persisting, would benefit from medical intervention, but we aren't yet able to predict who these ideal candidates will be. Good clinicians would arguably be in favor of approaching transition carefully and only after other possible avenues have been ruled out.

But when the majority of experts are too afraid to publicly criticize gender affirmative therapies, parents cannot trust that their child is being given a proper diagnosis. Based on the research regarding desistence, if a child's gender dysphoria persists into puberty, social and medical transitioning would be considered then.[8] But now, parents are rightly skeptical if an adolescent or young adult is given the go-ahead to transition, because clinicians aren't able to have the necessary conversations with their patients anymore.

For a child who would have desisted, transitioning amounts to a needlessly challenging process to undergo—and that's without considering the difficulties of transitioning back. Even a social transition back to one's birth sex can be emotionally difficult for children to undergo.

Many advocates for early transitioning will argue that a social transition is harmless, since it doesn't involve any medical interventions. In actuality, research has shown that socially transitioning is associated with persistence.[9]

The difficulties associated with detransitioning are illustrated in a 2011 study in *Clinical Child Psychology and Psychiatry* of twenty-five adolescents who had been gender dysphoric as children. Two girls, ages fifteen and eighteen,[10] who had undergone a social transition (taking on a male appearance) regretted it and detransitioned back to being female. Even though detransitioning from a social transition is fully reversible, it can be a difficult process, as illustrated by these girls, who feared teasing from their peers and waited until high school so that they could start over without anyone knowing about their past.

Watching this discussion unfold, my favorite justification for ignoring the desistence literature would have to be when an expert says, "We don't have enough research yet." A close second would be, "Little is known about these children," or "The existing research is flawed."

The denial of desistence requires us to forgo a conclusion backed by all of the available research, for a claim that has yet to be backed by any of it. In this climate, it's not hard to predict what new studies coming out of the woodwork are going to find.

It's also crucial to note a critical methodological flaw in their study design—the studies currently being undertaken do not include a control group, or a group of gender dysphoric children who are *not* undergoing a social or medical transition (or, alternatively, a different form of treatment, such as watchful waiting—described in greater detail below—or attempts to reconcile their birth sex). In today's climate, it is considered unethical to deny a gender dysphoric child the intervention being studied. To further remove potential bias, gender dysphoric children would also ideally be randomly assigned to either transition or not transition, but again, such an approach would be deemed unethical.

In these studies, all of the gender dysphoric kids will undergo treatment and are compared with siblings or same-aged peers who do not have gender dysphoria. Any similarities (or differences), such as good mental health, found between the two groups are then attributed to transitioning. Indeed, such benefits *could* be due to transitioning, but it's possible the same results would be found in children with gender dysphoria who *didn't* transition. Without this control group, we can't really know that attributing any such characteristics to transitioning is appropriate.

Political Pawns

There is no such thing as a transgender child. The stringing together of the words "transgender" and "child" is part of a lexicon pushed by activists to co-opt the young into their political movement. Many people, including parents, unknowingly use the term, not knowing there is a difference between being "transgender" and being "gender dysphoric."

"Transgender" is an identity and political label denoting that an individual identifies as the opposite sex. Children, and particularly prepubescent children, do not possess the emotional maturity to identify this way.

Gender dysphoria, as defined by the *DSM-5*, is a medical condition manifesting as distress at the incongruence experienced between one's experienced gender identity and one's birth sex. A child may be diagnosed with gender dysphoria.

Although the two terms are sometimes used interchangeably, more recently, a distinction has been made, but not for the reasons you'd expect. Mental health professionals now emphasize this distinction so that a person can identify as transgender *without* experiencing gender dysphoria.

The number of people who identify as transgender in the United States has doubled from three in 1,000 to six in 1,000 in the last ten years. It's possible that this considerable increase can be explained by greater social awareness and acceptance of trans people in addition to improved data collection methods, as researchers have described. But a quick look around suggests other factors may also be at play.

Transitioning, also known as "gender affirmative care" (which can take the form of social changes, like changing one's name and clothing, and medical interventions, like pubertal blockers, cross-sex hormones, and surgery), has become accepted by organizations like the American Academy of Pediatrics as the appropriate treatment approach for gender dysphoric kids.

These developments are in tandem with society's hailing of so-called "gender-creative" children, "gender explorers," and kids who are "gender-expansive." Supporting a young child's transition is seen as a beacon of tolerance, empathy, and love. Colleagues have told me stories of educators gleefully announcing when a child in their class has come out as trans, because they know that supporting that child in their new gender role will gain them attention and praise from their superiors.

This is no more apparent than in the flooding of news reports applauding parents for parading their transitioned children, including some who are preschool-aged, in front of a public audience, and politicians who are more than willing to pander and prostrate themselves in order to attract the vote of sympathetic Democrats.

According to the Human Rights Campaign, an LGBT+ advocacy and lobbying organization, there are close to fifty gender youth clinics in the United States. Britain's largest clinic devoted to treating gender dysphoric youth, the Tavistock and Portman NHS Foundation Trust, saw over 2,500 referrals in 2018–19, which included children as young as age three.[11] This was a 25-fold increase in the last decade, predominantly among girls transitioning to male.[12]

Can a child as young as three years old know their gender? As transitioning is being recommended at increasingly younger ages, what used to be considered a passing phase in a child's life has now become a monumental marker indicating that something must be rectified.

Calling into question aspects of transgender activism's narrative—that a child who says they are transgender should be supported in transitioning; that the sudden increase in adolescent girls transitioning is due to greater social acceptance of trans people; and that no one ever regrets transitioning, and if they do, they were never really trans—gets a person called all kinds of names, but this doesn't render the arguments untrue.

In the event that a child can grow to be comfortable in the body they were given, it shouldn't be controversial to contend that this would be a better outcome than a lifetime of hormones and possible surgery and sterility.

Transitioning is being sold as a normal part of growing pains and healthy psychological development. Parents and patients are told it is easily reversible, should someone change their mind—transitioning is the right solution for everyone, no matter how old they are, how long they've been feeling this way, or any other issues they are dealing with in their life.

In 2019, the World Health Organization removed "gender incongruence," its version of gender dysphoria, from being categorized as a mental disorder in the International Classification of Diseases, redefining it as a condition related to sexual health. This was presumably to alleviate the stigma that often comes along with having a mental disorder and to push back against those who dismiss the transgender community as mentally ill. The true problem, however, is not that gender dysphoria was considered a mental disorder, but that mental disorders are stigmatized. Reclassifying the condition fails to adequately address the source of this stigmatization.

Although doing so may indeed help to eliminate discrimination against people experiencing gender dysphoria, which is a good thing, the decision was shortsighted. Many compare the decision to the American Psychiatric Association's removal of being gay from its list of mental disorders in 1973. The comparison is not accurate. Gay people do not require medical interventions to alleviate their distress. If a gender dysphoric person is not experiencing a mental health disorder, why do they need to transition to feel better?

Transgender activists have gained much political ground through use of the narrative that being trans is similar to being gay—a person is born this way, it cannot be changed, and to question this is harmful to a trans person's well-being. The ironic thing is that many of these children are indeed gay, and as we'll see, by being persuaded to transition, they are actually undergoing a new form of conversion therapy.

The Medical Battleground

Psychotherapy, or talk therapy, aimed at children with gender dysphoria typically falls into one of three categories. The early transitioning or "gender-affirmative" approach is probably the best known because it has been touted by experts as the only ethical form of treatment, one that "saves lives." A child takes on a new name, changes their haircut and clothing, and is referred to using opposite-sex pronouns (although the prevalence of nonbinary children who are referred to as "they" is growing; see Chapter 8).

Everywhere we turn, we are told that gender dysphoric children "thrive" once they transition and their parents accept their true identity as the opposite sex. I would agree that parental acceptance is indeed important and that kids presenting with gender dysphoria, whatever the reason, deserve to be treated with compassion and unconditional love. This acceptance, however, does not require a social or medical transition.

Advocates claim that parents, teachers, and clinicians following the affirmative approach are merely supporting the child in choices around their appearance and toys, but this is misleading, because these aren't the only things a social transition consists of. Socially transitioning means that everyone who knows the child will refer to them as the opposite sex, endorsing the belief that they are *not* the sex they were born as. There is no reason, however, why adults can't support a child's preferences while maintaining that they are the sex they were born as, and—based on the research literature on desistence—at puberty, reconsider this, if the child's dysphoria persists.

A second approach, called "watchful waiting" or "wait and see," allows the child to guide the course they take. They may or may not eventually decide to transition.

The third approach, called the "therapeutic" approach or the "developmental model," allows a child to explore their gender while being open to the possibility that they may grow comfortable in their birth sex. A clinician will seek to understand factors relevant to the child's development, including trauma or other psychopathology, and what else is going on in the child's life that may be leading them to feel this way.

This final approach, which is backed by the scientific literature as the most appropriate course of therapy for these children, has instead been denounced by experts, medical organizations, and academic researchers as transphobic, "conversion therapy," and abusive.

Conversion therapy that aims to change sexual orientation is harmful because sexual orientation is fixed; it cannot be changed. But because gender identity is flexible in children, it isn't appropriate to conflate therapies helping them grow comfortable in their bodies with those that ineffectively seek to change sexual orientation.

I understand why such an approach makes people so uncomfortable. Efforts to change one's sexual orientation to heterosexual were once widely accepted. Thankfully, society has come around and most people see the error in that way of thinking. But with that understanding has come a resistance to any attempts at challenging an individual's experience or identity, particularly around gender and sexuality.

This means any kind of questioning of gender identity has been conflated with conversion therapy, and therapists and doctors are too afraid to suggest otherwise, because it will cost them their livelihood and reputation. For example, in the Canadian province of Ontario, Bill 77—the Affirming Sexual Orientation and Gender Identity Act—explicitly prohibits any treatment that attempts to change a minor's gender identity and delists any such treatment for adults from being covered by the province's health insurance plan.

A 2019 study showed that extended counseling can be effective in resolving adolescents' gender dysphoria.[13] Similarly, the Royal College of General Practitioners issued a position statement calling attention to the lack of evidence for the use of puberty blockers and cross-sex hormones, and saying that more research needs to be done regarding other treatment approaches, like "wait and see."[14]

If an expert defends research suggesting that other therapeutic approaches are more effective, they will be seen as unethical and operating outside of what is considered acceptable in the field. As a result, young children are being funneled through a series of chemical and surgical interventions that might have been entirely unnecessary had they been left alone.

The use of hormonal blockers to halt unwanted masculinization or feminization in gender dysphoric children is an off-label use. Puberty blockers have been approved by the Food and Drug Administration to treat precocious puberty, but we don't yet have any long-term data on their use in gender dysphoric children.

For these children, the onset of puberty is distressing because its associated changes make it more difficult for an individual to present as the opposite sex. In boys, unwanted effects include a growth spurt, facial and bodily hair, and the deepening of one's voice. Girls must contend with menstruation every month and the development of hips and breasts.

Parents are told that blockers are a "pause button" that will buy their children more time to decide about transitioning, and if they should stop taking them, puberty will resume as normal. I asked William Malone, a board-certified endocrinologist and the medical director at an endocrinology and diabetes center in Idaho, if this was true. He told me, "Despite the current widespread use of puberty blockers for gender dysphoria, the scientific community is largely ignorant to [these] answers."

He continued, "Puberty is a critical time for physical and psychosocial development, as sexual differentiation, brain, and musculoskeletal maturation all occur within complex social context. In other words, it's not just the physical changes of puberty (which are multifaceted) that are important, it's also the timing of those events. . . . We have no idea if fertility will be preserved, or if other physical functions of the body that depend on properly timed puberty will 'catch up.' "

As well, puberty blockers are associated with persistence. In one study from 2011, all seventy children who initiated puberty suppression went on to take cross-sex hormones.[15]

The 2011 Swedish study mentioned in Chapter 4 showed a twenty-fold increase in death by suicide for transgender individuals who transitioned medically and surgically in adulthood, compared with nontransgender people.[16] Some have interpreted these results as evidence that being forced to go through puberty leads trans people to have worse outcomes, but this can't be known.

What's interesting is how the findings of the same study have been molded to fit two different ideological positions on transitioning, depending on which side of the political aisle one is operating from. For those who are against transitioning, the increased suicides are a sign that transitioning doesn't help. For those in favor of transitioning, the increased suicides suggest that help didn't come soon enough. (There will be more on this when we discuss the suicide narrative parents are being told.)

The Endocrine Society's guidelines state that treatment is not recommended for prepubescent children. Puberty blockers are recommended at the start of puberty (between the ages of roughly 8–13 years old in girls and 10–14 years old in boys) until age 16, after which cross-sex hormones can begin. Surgery is recommended at age 18 or legal adulthood.[17] Some medical providers will make exceptions, however, allowing for surgery in children under the age of 18 if certain parameters are met, such as having two letters of support in addition to parental consent and being on cross-sex hormones for at least a year.

Malone voiced concern about the consequences of delaying a natural bodily process. "On its face, interrupting this intricate process that is vital to normal human development . . . in an effort to relieve psychological distress, is a risky proposition for which there should be significant scientific evidence. But evidence that it leads to more benefit than harm does not exist." Potential side effects include lowered spinal bone mass density* and, as shown in one study using an animal model, differences in spatial memory.[18]

According to numbers from the Tavistock Centre, between 2012 and 2018, 267 children under the age of fifteen began using puberty blockers. The BBC recently reported that a study being conducted at the clinic showed, after a year of being on blockers, an increase in reported suicidal ideation and self-harm.[19] (Of course, it isn't possible to infer causation from this study, since the study didn't include a control group of gender dysphoric children *not* taking blockers.)

Some might argue that these side effects are a small price to pay for a child's well-being. But by blocking puberty, a child is preventing the process that would have likely led to the resolution of their gender dysphoria.

In another study, the minimum age for the administration of cross-sex hormones was lowered from thirteen to eight years old.[20] Side effects in trans men taking testosterone can include an increased

---

*For example, see Klink, D., Caris, M., Heijboer, A., van Trotsenburg, M., & Rotteveel, J. (2015). Bone mass in young adulthood following gonadotropin-releasing hormone analog treatment and cross-sex hormone treatment in adolescents with gender dysphoria. *Journal of Clinical Endocrinology & Metabolism, 100,* E270–E275.

risk of heart attack, high blood pressure, diabetes, high cholesterol, and painful orgasm.

"Vaginal tissue is particularly vulnerable to the effects of estrogen deficiency, which in this case mimics a post-menopausal state. Symptoms of estrogen deficiency could include vaginal inflammation, dryness and pain, and sometimes urinary incontinence," Malone said.

For trans women taking exogenous estrogen, they are at an increased risk for blood clots, stroke, and cancer.

In my earlier conversation with Buck Angel, he told me he nearly died due to the absence of knowledge regarding transgender medicine. After hormonally transitioning, his uterus had atrophied, fusing with his cervix, causing intense cramping. Health professionals dismissed his concerns on multiple occasions. The infection eventually burst and became septic.

"I transitioned 23 years ago. I'm a guinea pig. I have had a lot of things happen to me. One of the most dangerous things . . . was atrophy, yet we don't even talk about it."

The path of medical intervention is one that must be seriously contemplated before being embarked upon, because it is lifelong. Not all transgender people will elect to undergo a full medical transition, but if it is what an adult decides is best for themselves, we should be on board, no matter how challenging it may be.

For children, however, the reality of desistence and the misuse of suicide statistics call for a moment of reconsideration.

A Happy Daughter or a Dead Son?

At the heart of the childhood transitioning debate is a mantra that should make every mental health professional recoil. I can't count the number of times I've heard a parent recite the words, "Do you want a happy daughter or a dead son?" (or "Do you want a happy son or a dead daughter?") as both something a clinician once asked them and the reason why they allowed their child to transition. The earliest documented use of the phrase appeared in an article in 2011 by ABC News, in which one of the leading researchers in the field describes how she often asks the parents of her patients, "Would you rather have a dead son than a live daughter?" before referencing the transgender suicide rate. Since then, the mantra has been repeated ad nauseam by experts and parents in media coverage about childhood transitioning.

Asking any parent whether they would prefer a happy, flourishing child or one that is dead is morally bankrupt. What kind of mother or father would say they'd prefer a dead child?! This is how unknowing parents, who only want to make life easier for their son or daughter, have become prime targets of a campaign that has weaponized science and medicine and taken vulnerable children as prisoners.

A related form of emotional blackmail involves constantly referencing the suicide statistic, which has little relevance to these children. A study from 2014 found that 41 percent of adults who identify as transgender have attempted suicide at some point in their life,[21] but the researchers of the study acknowledged its limitations. They didn't ask respondents about comorbidity, such as other mental health conditions, nor whether they identified as transgender at the time of the suicide attempt. It is possible that the suicide attempts reported by individuals in this study had nothing to do with how they felt about their gender.

Parents across the board are consequently being encouraged to monitor for signs of gender-atypical behavior in their children. Those parenting gender dysphoric kids are also irresponsibly being told that 41 percent of transgender *children* have attempted suicide (even though children under eighteen were not part of the study) and that their child has a high likelihood of becoming part of this statistic.

The misuse of this statistic is dishonest and cruel. But it is effective, which is why it continues to spread. Parents who wouldn't otherwise allow their kids to make basic decisions about what they're having for dinner or what time they're going to bed will support a child's wish to undergo potentially irreversible medical interventions. Critics of the "trans kids" movement will frequently accuse these parents of being irresponsible and unfit to parent, but that's not necessarily what it's about. Logical, otherwise wary parents are having their deepest, most natural fear exploited by being told this is what their child needs to survive.

Another part of the problem is that, regardless of political affiliation, no one is incentivized to promote a statistic that is less shocking. On one side, it feeds into scaremongering tactics for those in favor of transgender rights, and on the other side, it's a justification for those who believe gender dysphoria should be treated as purely a mental phenomenon and we should not be promoting transitioning for anyone.

What I've learned is that parents who are pro-transitioning are heterogeneous. Although the ones who receive media coverage seem fully confident in their decision, this is not the case for every person parenting a gender dysphoric child. Some are fearful for their child, but believe they don't have any other choice. Disagreements between parents are playing out in an increasing number of custody battles in which one parent supports a child's transition and the other does not.

Returning to the father I spoke to at the start of this chapter, I was surprised that he wanted to speak with me. I would have expected that most parents of children who had transitioned would rather cuss me out or have me exiled to another planet. My work, despite not being directed at them personally, essentially said that they were endorsing the wrong approach for their child.

As for children, and particularly prepubescent children, clinicians and researchers in the field have been bullied and intimidated into silence, leading to policies and therapeutic practice that does not promote the best outcomes for gender dysphoric youth. For example, in 2019, Marcus Evans, one of the governors of the Tavistock and Portman NHS Foundation Trust, resigned after citing concerns about its approach to treating gender dysphoric children. In his resignation email, he stated that attempts to discuss the issue from different perspectives are being impeded by accusations of transphobia and prejudice.[22] For its part, the Trust said it "did not identify any immediate issues" regarding the safety of its patients, but was developing a plan of action for identified areas of improvement.

The hypocrisy of transgender activism is palpable. The same people who will argue (as we should) against involuntary intersex surgeries, based on the understanding that young children do not possess the mental capacity to consent to invasive surgeries, will argue that an eight-year-old has the emotional maturity to begin the medical process of identifying as the opposite sex. Physicians won't even allow adult women to make a decision about tying their tubes or having a hysterectomy until they are in their thirties.

Transgender activists pushing this agenda are the vocal minority, and although they wield immense influence, they don't reflect everyone in the community. I've had many transgender people reach out to me over the years, saying they agree with me and are similarly critical of children transitioning.

Nevertheless, this movement has taken on a life of its own, only managing to gain further momentum, the more time that goes by.

At a time when criticism of any of these ideas is shot down as transphobic concern-trolling and dog-whistling to anti-trans groups, some of you may be appalled at my criticism, thinking, "Do we really have the right to question the experiences of trans people?"

I say we do, particularly if refusing to do so causes harm to others. We can be in favor of equal rights for the transgender community, which is something I fully support, without bending unpopular truths.

I, personally, would have never thought I'd be on the side urging for caution against young children transitioning. I was previously in support of the idea because it superficially made sense to me—why not prevent the irreversible effects of puberty that would otherwise cause these children unnecessary grief and emotional harm? (Not to mention the skyrocketing suicide rates, which I also believed.) Before becoming a sex researcher, I too watched the uplifting media stories of children living as the opposite sex and believed this was a positive thing. My opinion changed once I read the research literature on desistence and realized the public has not been privy to all of the relevant information.

As a result of saying this aloud, I've been accused of pitting the L's, the G's, and the B's against the T's. I'm constantly amazed at the number of gay men who will publicly defend childhood transitioning when the movement is in fact leading to the extermination of gay children.

As would be predicted by the desistence statistic—which, let's not forget, is based on more than four decades of research—the vast majority of children who voice the desire to be the opposite sex will eventually change their minds, growing up to be gay and comfortable in the body they were given.

It's possible that these men were gender-conforming and masculine as children, and as a result, fail to make the connection between gender dysphoria in childhood and being gay as an adult. But I'm willing to bet it's more likely they've forgotten how much they were once like these kids. Research has shown that grown men who experienced gender dysphoria as children will forget they ever felt this way or that they once voiced the desire to be a girl.[23]

Behind the scenes, I've heard many stories of how the gay community is terrified of criticizing transgender ideology, because gay men who are not ethnic minorities are now considered to be white men with privilege. (According to identity politics, any oppressed group that has managed to overcome adversity gets lumped in as "white men," who are seen as holding all of the power in society. Gay white men are in good company, however, because the same logic has also extended to Jewish people and Asians.)

I understand the reasoning behind this narrative. Transgender people face harassment and discrimination for simply being who they are, and many supporters probably genuinely believe this is the right choice for these children.

What it all comes down to is that the information parents are being given is lopsided. They are being told that anything but affirmation will lead a child to attempt suicide. The effects of this narrative run deep, and have everything to do with the growing phenomenon of rapid-onset gender dysphoria.

Rapid-Onset Gender Dysphoria

In Chapter 4, we discussed how, for some children with gender dysphoria, the fact that they are different from their peers is apparent from the moment they are born. These kids prefer to dress and behave like the opposite sex and have mostly cross-sex friendships.

For other children, they don't begin experiencing issues with their gender until adolescence. Over the last ten years, there has been an abrupt increase in the number of gender clinic referrals for adolescent girls wanting to become boys. The sex ratio for referrals, once dominated by males, has since inverted.[24]

According to the Centers for Disease Control and Prevention, approximately 2 percent of high school students identify as transgender.[25] Progressives have attributed the sudden explosion of girls coming out as transgender to greater visibility of the trans community and the belief that better awareness and availability of support has led to a society-wide reduction in transphobia. But if this were the case, one would expect an equal number of individuals of both sexes to be coming out with a new gender identity.

Until recently, adolescent-onset gender dysphoria had only appeared in adolescent males (in the form of autogynephilia). Rapid-onset gender dysphoria (ROGD), seen primarily in teenage girls and college-aged young women, is characterized by a sudden desire to transition to male, often out of the blue, without any previous history of gender dysphoria. Similar to desistence, this issue has been called a "myth" by some critics, a condition made up by transphobic parents and researchers, subsequently fueling backlash against those who study it.

In August 2018, all hell broke loose when Lisa Littman, a physician and assistant professor at Brown University, published the first study on rapid-onset gender dysphoria, which consisted of 256 parents completing a ninety-question survey. Her findings suggested that for some of these girls, being transgender is a form of social contagion.

More than 25 percent of the children had come out as gay, and more than a third had come out as bisexual, prior to identifying as transgender. About 60 percent of the children had at least one mental health disorder, such as anxiety or autism, and many others had a history of self-harm or trauma. From the conversations I've had with these parents, many of the girls also have a diagnosis of borderline personality disorder. Most important, as noted in Littman's study, none met the diagnostic criteria for gender dysphoria in childhood as defined by the latest *DSM* guidelines.

Regarding the social contagion factor, there was an association between suddenly coming out as transgender and having a friend (or multiple friends) who also identified as trans. For about 40 percent of these adolescents, more than half of their friend groups had also come out as transgender. This is more than seventy times the prevalence of transgender adults in the general population.

The desire to transition typically emerges after an individual has spent much time researching gender dysphoria online or watching transitioning videos on social media, wherein puberty blockers and cross-sex hormones are glamorized. Some of the Internet forums accessed by the children even provided instructions on how to get approved for hormone therapy. The attention and validation they received from coming out as transgender also proved to be very powerful. The study showed some of the potential social benefits of coming out as trans, including increased popularity among peers when they were previously ostracized, and greater protection by teachers from bullying, because teachers were more concerned about anti-trans bullying than bullying that targeted gay students.

Due to activist pressure and public outrage, the study was quickly denounced as "transphobic." Some left-leaning media publications called it "junk science" and "anti-trans," and alleged that the parents who took part were recruited from right-wing hate groups.

In response to the backlash, Brown University pulled its press release for the study, and the scientific journal that published it announced it would be conducting a post-publication review of its content and methodology.[26]

Reconsidering the methodology of a peer-reviewed study after publication is essentially unheard-of. What was most disturbing was that opponents of Littman's study clearly had not even read it. Roughly 86 percent of the parents who completed the survey endorsed marriage equality and 88 percent supported transgender rights. (Littman described herself to me as a liberal Democrat.)

These parents were not anti-trans by any means. Many have told me, if their child was transgender or gay, they would fully support them. Their hesitation stemmed from something not quite sitting right—the rashness of their child's desire to transition seemed incongruent with everything they knew about their previously happy-to-be-gender-typical child.

For those who are on the autism spectrum, wanting to transition may be motivated by rigid ideas about gender roles. A child who harbors a disdain for dresses may interpret this as a sign that she should really be a boy. Transitioning could also be a form of a highly focused and intense interest, which is characteristic of being on the spectrum.

In the words of Susan Bradley, an international expert on gender dysphoria in children, a professor emeritus in the Department of Psychiatry at the University of Toronto, and the former psychiatrist in chief at Toronto's Hospital for Sick Children, "When I start identifying these traits in these adolescents who present with gender dysphoria, I really want to be clear whether they've had a previous history of having what we call 'repetitive and obsessive interest' in certain things. Because once some of these kids fix on an idea, they can become really quite obsessed by [it].

"I've had some parents who, once I've asked about all of this, they'll say, 'Well, sure, they were into [action figures], and they were into a variety of other things, but it went away after a while.' That's part of what we need to understand, and also help them understand. Is this something that is a little bit like some of the things you get obsessed with, or is this something that's going to be necessary for the rest of your life? And when you look at the stories of some of the people who have changed their minds later on, a lot of them are saying very strongly, 'I wish somebody had understood me from a psychological point of view and didn't just take at face value that I thought this was the answer to how I was feeling.' "[27]

In many cases, an individual is attempting to cope with prior sexual trauma or sexual abuse and wants to be rid of sexualization by society. Bradley continued, "Some of these girls, before they are transitioning, have been sexually abused or threatened. They feel that they're unable to protect themselves as females and that's another spur to say, 'I would be better off as a male.'

"I saw a young person many years ago who came to the clinic with an expressed wish to be male. And after we'd done our interview, it was really clear, she had been sexually abused by a couple of boys and she felt horrible about it and she started to put on weight and started to change her appearance so that she was bigger and stronger. And she said, 'Nobody would bother me now.' "

One of the most prominent criticisms of the study was that transgender people weren't consulted in its implementation. This complaint is foolish, however, because the role of a scientist is to be objective. Whether or not they identify as part of the population or community they are studying shouldn't make a difference, and a properly designed study, along with institutional review boards to ensure research is conducted in an ethical manner, operate with these concerns in mind. Considering that Littman paid for the study out-of-pocket and does not earn her livelihood providing services related to transitioning, she had fewer conflicts of interest than many of the other researchers studying this subject.

It was unusual for a scientific journal to place an already published paper under review for a second time, considering it would have been evaluated by other experts in the field as part of the peer-review process prior to publication. In 2019, the journal in question republished Littman's study, noting a "correction." The correction was flaunted by left-leaning media as a sign that justice had been served. In actuality, the correction consisted of giving the study a new title and rewording its language in a few places. There was nothing wrong with its original analyses or findings.

To be fair, I can understand why talking about ROGD can be seen as threatening to the transgender community and its allies, as the discrimination and hardships that trans people deal with are unfortunately prevalent. As I've said before, I believe transitioning can be the right answer for some adults, but the same cannot always be said for children, particularly those whose desire to transition is rooted in something other than unhappiness around their gender. There are any number of reasons why a young woman may feel uncomfortable or dislike being female; they don't necessarily mean that transitioning is the right solution for her. With ROGD, a girl's proclamations of gender dysphoria usually have nothing to do with gender.

What I found the most unnerving when I first began looking into ROGD is the degree to which each child's story was identical. A previously feminine daughter would attend a school assembly about transgender issues or perhaps befriend an influential peer who later comes out as trans. The child will then socially transition at school without their parents' knowledge. When the parents are finally alerted to this new identity, they are told the child will be at a high risk of suicide unless their gender is affirmed.

Taking a child to a gender clinic only makes things worse. As an indication of the pace some gender specialists choose to move at, I have had parents tell me that, upon calling a gender clinic to simply gather information, the receptionist would tell them, without ever having met their daughter, that the child must begin hormonal blockers immediately.

This speed is amplified in parents who have gotten on board with their child's transition. At an event several years ago, a woman approached me to tell me her story.

"My daughter—I mean, my son," she said, "has come out as transgender."

She explained to me how her fifteen-year-old child, who was always very girly, told her one day that she wanted to become a boy. The mother, initially hesitant, took the child to a gender clinic, where she learned about the high rate of suicide among transgender teens and realized the best thing she could do was support her child.

"He has started binding and he wants to start testosterone, too," she said. "I asked him, 'Are you sure you don't want kids?' He says he doesn't. But at that age, I didn't, either."

"Is your son seeing a therapist?" I asked, seeking at least a glimmer of hope that someone might pick up on something being amiss.

She nodded, praising the group therapy he had been attending. "The kids in the group, they've transitioned and seem to be doing much better," she said.

I didn't know how to tell her that yes, of course, the kids attending a therapy group were probably finding the group useful; otherwise, they wouldn't be there. This said nothing about the kids who didn't find it helpful and stopped going, or those who decided not to pursue transitioning.

For those who counter evidence supporting ROGD with the belief that these girls were transgender all along, managing to hide this out of fears of parental rejection, it isn't difficult for a caregiver to pick up on cross-sex interests and behavior, especially if they have known the child from birth. Some of these girls will profess a love of makeup and feminine clothing as little as a week before announcing they are transgender.

In the end, a legitimate scientific study was publicly undermined as the result of a single activist complaining on social media. I can count on one hand the number of sex researchers who publicly voiced solidarity with Littman's study. The entire fiasco was a reminder of what happens when activists become enraged.

There will be real consequences to this silence, including, as we'll see, the phenomenon of detransitioning. Since legitimate experts now avoid anything to do with treating gender dysphoria in children, activists have stepped in, taking their place. "Gender experts" may or may not have any knowledge about the proper course of action for gender dysphoric kids, but they are willing to refer patients to other professionals who will facilitate what they want.

Bradley agreed: "A lot of people have said that they're afraid to speak out about it. The other clinics, they have appeared, at least to me, to accept at face value an individual's statement rather than doing what we would regard as a proper psychiatric and psychological assessment of the child and family." This is because they're afraid of losing their jobs.

Worryingly, the conversation now revolves around transitioning as opposed to whether someone experiences gender dysphoria. Transitioning is no longer a medical consideration; it has become an identity. A cohort of younger millennials and Gen Z have become the test subjects of this ideology.

"There's a lot of butch women who are transitioning now at a rapid rate," Angel told me. "Because I have heard from some, saying that it's easier to live as a man than a woman. This is what I've heard from actual people transitioning. If you want to transition for that reason, go right ahead, but then we're missing the whole point of gender dysphoria."

ROGD speaks to the fact that it's become more socially acceptable to be a trans man than a lesbian woman in certain peer groups and progressive circles. For women, regardless of sexual orientation, the strength and power associated with being a man is alluring. For example, men do not face the double standard of being called names when they are assertive.

Moving forward, it's critical that transgender activists feel they can engage constructively with research scientists in order to have their voices heard. Shutting down this research is counterproductive, as doing so only adds fuel to the fire for those who seek to delegitimize transgender people. It also derails our ability to successfully help those suffering from gender dysphoria.

Regarding advice for parents, Bradley suggested, "The parent has to ask questions like, 'How long have you been [feeling] that way? What makes you feel that way?' You can't shut down the conversation. You have to keep the conversation open."

As for young people who may be questioning their gender, Bradley said, "My advice is that it's really important that you understand why you feel this way, especially if you have felt this way fairly suddenly or recently, as opposed to having had a long-standing feeling that you were not the right gender for your body. But it's really important that you don't get into doing something that may be harmful to your body until you're absolutely certain this is the only path or the best path for you, and that may take quite a long time to work out." She suggested seeing a therapist who is broad-minded, or attending group therapy providing a variety of differing points of view, including talking to gay adults who experienced gender dysphoria in their youth.

The Role of Homophobia

Despite the many strides forward made by the gay rights movement, children are being encouraged to transition as a solution to society's homophobia. Parents undoubtedly notice if a child is gender-atypical. Effeminate sons tend to evoke a stronger negative response than tomboy daughters. With greater public awareness about gender dysphoric children and the difficulties and stigma that they face, parents will receive more admiration and support when raising a transgender child than a child who is gay. For those who are troubled at the thought of having a son who is girly, transitioning offers a promising solution—by allowing a feminine boy to transition, he now presents as a feminine girl. A little boy who enjoys playing with makeup and dolls with his female friends will bring about much less attention and criticism if he, himself, is a girl.

This same logic extends to a child's sexual orientation and their future romantic partners. On some level, these parents know there is a chance their feminine son will grow up to be sexually attracted to men (and conversely, that a masculine daughter will grow up to be attracted to women). Instead of allowing this to happen, they will be more than happy to facilitate their child's requests to transition so that, to the outside world, that child will appear heterosexual. An adolescent boy who is attracted to other boys will, upon transitioning to female, appear straight.

What is most disturbing is that these parents will be praised as open-minded, examples of what true love and acceptance looks like, "on the right side of history," when in actuality, some are homophobic[28] and endorsing a repackaged, socially acceptable form of conversion therapy. In some cases, a child may internalize their families' antigay sentiments, which will add to their desire to transition.

What I want to know is, what's wrong with having a feminine son or a masculine daughter? Assuming a feminine boy or a masculine girl must *really* be the opposite sex is supportive of gender stereotypes and backward in thinking. The same line of reasoning extends to our consideration of gender-atypical people to be a different category altogether, like "gender nonbinary" or "genderqueer" (see Chapter 3), instead of expanding the definition of what it means to be a man or woman.

A related theme is the posthumous revising of prominent historical gay figures. For example, Marsha P. Johnson, an American gay rights activist who passed away in 1992, was vocal about being a gay man and a drag queen, but is frequently cited today as being a "trans woman of color."[29] Lesbian erasure is also evident when the efforts of lesbian women are rewritten as belonging to members of the transgender community. The Stonewall riots, which marked the first uprising of the gay community against police interference, were started not by trans women, as is commonly stated, but by Stormé DeLarverie, a biracial lesbian.[30]

Since I began writing about this issue several years ago, countless gay men have reached out to me, thanking me for speaking out about this. Many of my friends—including some whom I had not spoken to in years—have told me how relieved they are to have not grown up in today's political climate. As children, they similarly voiced unhappiness about their bodies and believed that they should have been girls, but eventually grew up to feel comfortable living as gay men. They are grateful that transitioning wasn't as widely promoted, because they were certain they, too, would have been encouraged to transition by the adults in their lives.

Many young lesbians are dissuaded from identifying as lesbian—or as women, full stop—because they do not wish to be hypersexualized. This is in part due to the association with pornography created for straight male audiences. (According to recent statistics, "lesbian"-themed pornography ranks as one of the most popular search terms.[31])

Society is unkind to masculine women, and particularly butch lesbians. Young lesbians who have detransitioned will often say they do not feel they have lesbian role models to look up to, and that they experience internalized homophobia and shame for not being conventionally feminine. When we consider public figures in the media, most women who are sexually attracted to women now identify as transgender men, nonbinary, or genderfluid. Most of the lesbian representation we do see is transgender women who are attracted to women.

Detransitioners and Transition Regret

Detransitioners are individuals who once identified as transgender but have returned to identifying as their birth sex. The extent to which they transitioned varies; some went as far as a social transition before changing their mind, while others underwent hormonal treatment, and in some cases, surgery. Although male detransitioners exist, with the rise in rapid-onset gender dysphoria as a global phenomenon, the majority of individuals detransitioning more recently were born female.

Like many uncomfortable truths that call into question transgender ideology, transition regret has been branded a myth and statistically rare. Detransitioners have been swept under the rug and ignored, due to an ongoing narrative that their gender dysphoria wasn't real and they were never really transgender.

According to one Swedish study, an estimated 2.2 percent of people who transition subsequently regret it.[32] The true percentage is probably much higher, however, considering that this number pertains only to those who applied to legally change their sex, as not everyone who transitions and detransitions goes through the process of legally documenting both. In addition, study data were only collected until 2011, which was prior to the dramatic increase in transitioning rates, particularly in adolescent girls.

More recent reportage suggests the numbers are indeed much higher. In Newcastle, England, a city that has a population of about 300,000 people, hundreds of young women have come forward, saying they regret transitioning.[33] Fitting the profile of girls with ROGD, most report being lesbian and on the autism spectrum.

Why aren't we hearing more about this? One reason detransitioning seems rare is that a person isn't required to go to the doctor in order to do it. This makes it difficult for gender clinics to track these numbers. In most cases, over the course of several months, an individual will make the decision to halt medicalization by stopping hormones and socially transitioning back to their birth sex. Unlike the decision to come out as trans, which is often done publicly to alert friends and family on social media, the decision to detransition is usually undertaken quietly and privately. Some will disappear for a period of time before reemerging on social media as their birth sex without any formal announcement.

Transitioning is being sold as reversible, and in the event that someone changes their mind, that no harm will be done. In turn, detransitioning is viewed as medically and surgically going backward, simply undoing the changes associated with transitioning—but in reality, such a thing is impossible.

Testosterone leaves a woman's voice forever altered and gives her permanent facial hair, even after she has stopped taking it. For those who have had a double mastectomy or a phalloplasty (or, in the case of someone born male, a breast augmentation and vaginoplasty), they may choose to pursue reversal surgery through reconstruction of their breasts (or chest) and genitalia. In a 2018 study in *Plastic and Reconstructive Surgery*, 46 surgeons across two transgender health conferences reported that, of the approximately 22,725 transgender patients they had surgically treated, 62 were for transition regret. This translates to a detransition rate of about 0.3 percent; however, these numbers were collected in 2016 and 2017. We have yet to see the fallout from the spike in referral numbers resulting from ROGD.

In cases where an individual has had a hysterectomy (including removal of their ovaries), their only option is to be taking exogenous estrogen for the rest of their life. Similarly, for those born male who have had their testes removed, they will need to be on testosterone, and if they were taking estrogen, they will have permanently enlarged breast tissue.

In the event that an individual regrets their bottom surgery, it can be reversed, to some extent. For males who transitioned to female and back to male, a phalloplasty can be undertaken to create a new penis, but it will require the implantation of a rod to facilitate an erection for intercourse. These procedures are also expensive, ranging in the tens of thousands of dollars, not to mention the associated risk for complications that accompanies invasive surgery. Fewer females who have undergone a phalloplasty have requested a reversal surgery, so less is known about them.

Many detransitioners have been traumatized by surgery and doctors, so the last thing they want to do is undergo more surgery. Most just want to resemble their birth sex enough so that they can get by in society without any problems and they can move on with their lives. For those who have had a double mastectomy, insurance won't cover breast reconstruction, and most are still paying off the cost of their original surgeries. Even with implants, they may not be able to breastfeed a child.

Medical professionals are following direction from activist organizations. They are under the impression that if they don't offer access to hormones, no matter how young or to what extent psychiatric comorbidity is evident, an individual will commit suicide. They do not have any protocols in place to help detransitioners, because they see detransitioning as a problem within the LGBT+ community that should be solved among its members. They are also very likely motivated by fears of being called transphobic, because helping a patient detransition can lead to clinicians losing their license. With the passing of bills that outlaw any approach that isn't affirmation, clinicians have told me they are terrified to step outside the lines.

For those seeking to detransition, it can be a very isolating experience. One detransitioner I spoke with had to change her entire medical team after realizing that the doctors who helped put her on the path to transition weren't going to help her go back. Some detransitioners find solace in attending groups for cancer treatment survivors because they have also had organs removed, as well as groups for those who have experienced traumatic amputations, because they have nowhere else to turn.

Detransitioners are assumed to be conservative or right-wing, but this is not usually so. Upon detransitioning, most identify as gender-critical or radical feminists, and in many cases, they first encountered gender ideology, which led to their transition, in far-left progressive content on social media sites.

Across the conversations I've had with detransitioners, there is a common thread among their stories. Many experienced being misgendered early on in life, due to appearing masculine. Prior to transitioning, some experienced harassment in women-only spaces by women who thought they were men, and so they assumed transitioning to male would lead to an easier life.

The process of detransitioning usually begins with the realization that a person wasn't any happier post-transition. In many cases, they were actually less functional than before because their underlying problems were never addressed—whether it was a discomfort with being gay, disliking women's roles in society, or, as discussed earlier, comorbid psychopathology or a history of sexual abuse.

This is not to say that all transgender men would have been better off not transitioning, or that life is a walk in the park once they do. There are also very real fears from the transgender community that acknowledging detransitioners' experiences will lead to denial of medical treatment for everyone. This is understandable, considering how difficult it's been for the community to have its concerns taken seriously, because acknowledging that some trans people return to their birth sex has been used to support the argument that no one should be allowed to transition, ever. But what happens to those for whom transitioning was not the right choice? They've been left to their own devices, to suffer silently. Empathy for one group does not require disparagement of the other.

I would argue that detransitioners' coming forward doesn't mean all transgender people will change their minds. If medical professionals were able to do their jobs properly and rule out other possible explanations for a person's gender dysphoria, the vast majority of detransitioners probably wouldn't have transitioned in the first place.

Not all transgender people deny that detransitioners exist, however. When I spoke with Buck Angel, he told me, "For five years, I've been watching detransitioners, and I've been screaming, 'Hey kids, look at this!' and nobody will listen to me. Everybody slides it under the table, and in the trans world, you're not allowed to talk about it."

He said that young people frequently reach out to him, expressing confusion and disappointment about their path to medicalization. "Kids are saying they're not trans [and] why did they do it? They wish they had a doctor telling them. Why aren't we listening to this and saying, 'Wait a minute, community.' "

One detransitioner I interviewed, a twenty-two-year-old woman, first identified as nonbinary, then as a transgender man, during her teenage years. She took testosterone for over a year before detransitioning at nineteen. She said her decision to identify as transgender stemmed from feeling "fundamentally different" from other girls who were "organized," "soft," and "graceful," when she was, in her eyes, "messy," "loud," and "clumsy," in addition to battling an eating disorder.

" 'Being trans' on Tumblr was a fun distraction from the misery of my daily life," she explained. "I still had next to no friends at my new school, I still had a bad relationship with my parents, I still hated my fat, but at least I could enter this alternate universe where I was a cool guy with lots of followers. It started being a lot less fun when I began pursuing transition in real life."

Upon turning eighteen, and without her parents' knowledge, she drove six hours to the next city over to get access to testosterone. "[I] was prescribed testosterone . . . after one, one-hour appointment," she said.

She soon realized that transitioning was not the answer, and that it was in fact making her feel worse. She described funneling her anger and grief toward promoting trans activism on campus, securing a position on the executive board of her university's LGBT+ organization and attending protests. She jokingly added, "And persecuting professors who committed heretical crimes such as refusing to accept singular 'they' pronouns."

Her mental health, however, continued to decline. She began abusing substances and self-harming, eventually leaving school. "Needless to say, the whole thing about transition being 'lifesaving' and making everything better was starting to not make sense in my mind."

It's not unusual for a young woman to feel this way—discomfort at having a female body and feeling as though she is not like other girls. I certainly felt that way growing up. But instead of reassuring young women that this is completely okay and there is no "correct" way to be a woman, society is now convincing girls who are different that they are really men or another gender.

A question that is commonly asked is whether detransitioners were ever really transgender. Just because someone detransitioned doesn't mean they never experienced gender dysphoria or that their feelings weren't real. Similar to desisters, because detransitioners don't fit the story that trans activists would prefer to promote, they are dismissed by the community and told their dysphoria wasn't that severe. A more productive conversation would involve acknowledging these realities as evidence that transitioning is not the one-size-fits-all solution for everyone, while maintaining that this doesn't mean it benefits no one.

Those who only underwent a social transition before detransitioning are cast aside as "not trans," but shouldn't the same logic also extend to prepubescent children—and also grown adults—who have socially transitioned? Why is the trans-ness of a socially transitioned three-year-old given more weight than the former identity of an adolescent? If an adolescent who socially transitioned was "never really trans," then the same should apply to anyone who undergoes only a social, and not a medical, change.

Detransitioners also face accusations that they are in the same boat as ex-gays, individuals who are gay but claim they have been cured of any attraction to those who share their sex. This is not an apt comparison, though, because those claiming to be formerly gay are in denial of who they are. (As mentioned in Chapter 4, sexual orientation is biological and fixed.) Detransitioners are not denying their true selves; if anything, detransitioning allowed them to get closer to this reality.

Some people detransition for reasons unrelated to whether transitioning was helpful. Some do so out of an inability to access or afford treatment, including hormones, or in some cases because of the discrimination they encounter and fears for their physical safety.

Separating themselves from the ideology can be as difficult as detransitioning from the physical changes because it is ingrained in their identity and their social support is wrapped up around it. It isn't uncommon for a detransitioned person to lose all of their social support upon disclosing that they are detransitioning, and then be blocked on social media by their friends. They will be accused of having internalized transphobia and be told not to detransition, a common theme among the detransitioners I spoke with.

It is particularly disturbing that vulnerable young girls are being inculcated with this ideology, and that attempts to question or break free from it induces a second course of brainwashing. Across the many stories I've heard, the process of detransitioning, and living with the remaining side effects of transitioning, is particularly painful because one must face the realization that they did this to themselves.

As for the parents of detransitioners who signed off on treatment, many feel guilty. They believe they failed their child. Most sensed transitioning was not the right fit, but they overrode their parental instincts and listened to the doctor. For children who transitioned as minors, they must contend with the fact that their parents were on board with the initial decision. In some cases, a parent won't accept the choice to detransition because it goes against what the doctor advised, adding another layer of complexity to the child's decision.

The number of young women who are now regretting transitioning speaks to an absolute failure of the medical system, the negligence in avoiding this predicament in the first place, and now, the inability to help these individuals heal. I foresee many class action lawsuits in the future as more and more young people detransition and ask themselves how the adults around them agreed to it.

Personally, I cannot believe this is happening. I have spent countless hours in discussions with colleagues, shaking my head, hoping that we're wrong. The entire time I spent researching detransitioning for this book, I was stunned and speechless.

Transgender ideology has allowed young women the option to opt out of womanhood and having a female body. And when you are young and don't feel like you fit in, why wouldn't you? Society instructs women to be like men in every facet of their lives, but for women who are naturally masculine, they are told there is something wrong with them.

Within sexology, we saw this tragic period coming for years, the only logical outcome following a generation of children being rushed to transition without critical thought. We tried to stop the epidemic that is coming.* No one would listen.

There are few journalistic issues that affect me on an emotional level. As a scientist, I was taught to distance myself from my feelings; acknowledging them feels unprofessional.

The more I learn about detransitioners, the more heartbroken I become. There is no question in my mind that the 2 percent statistic of those who regret transitioning is going to multiply vastly in the years to come. When society looks back on this in horror—we tried to warn you.

*One prominent sex researcher noted how, in the span of one year, four sets of parents had brought their gender dysphoric child to see him after another gender specialist had started the child on hormonal therapy after only one visit. See Levine, S. B. (2018). Ethical concerns about emerging treatment paradigms for gender dysphoria. *Journal of Sex & Marital Therapy, 44,* 29–44.

MYTH #6

NO DIFFERENCES EXIST BETWEEN TRANS WOMEN AND WOMEN WHO WERE BORN WOMEN

"This is going to be so uncomfortable," I said to myself, as I set up the videoconference on my computer and cleared my desk.

For pretty much every other journalistic interview I've gone into, my biggest concern has been that the audio recording software worked as expected and my battery pack didn't run out. But on this day, I was interviewing a transgender activist I'd repeatedly butted heads with on social media and wasn't fully convinced that two minutes into our conversation we wouldn't be going off the rails.

When you work in media, you find a way to compartmentalize and ignore the online hate. But here I was now, about to face it quite literally head-on. To my knowledge, this particular activist had never written anything horrific or libelous about me, but in my mind, she was the physical embodiment of what trans activists thought about those of us who followed the mantra "Trans women are women" with "Yes, but . . ."

If you follow the culture war long enough, sooner or later, you'll encounter the word "TERF." TERF stands for "trans-exclusionary radical feminist," but its application has extended beyond its original definition to mean anyone who says, "Wait a minute—" to transgender credo, particularly in the context of pointing out how transgender women are different from women who are not transgender. The term first came about in 2008 as a way for radical feminists who were in favor of sharing women-only spaces with trans women to differentiate themselves from radical feminists who weren't. Like many things on the Internet, the acronym took on a life of its own and has since become a staple in the vernacular of transgender activists, essentially conveying the same meaning as "transphobic," regardless of whether one identifies as a radical feminist or a feminist at all.

I've been called a TERF more times than is humanly possible, even though I don't consider myself to be a radical feminist (and I'm sure most radfems would tell you I am not one of them). Chapter 2 of this book will surely confirm this, in the event that you have any doubt.

I'll explain how radical feminism fits into all of this in a moment. But as for the interview I was about to commence, I took a deep breath and clicked on the little icon to start the call.

After witnessing a long line of my sex research colleagues being harassed, abused, and threatened for decades—in some cases, having their livelihoods close to destroyed—by unscrupulous trans activists punishing them for their work, I learned that the only way to win this battle was to show no mercy.

Much to my amazement, the interview went a million times better than I could have anticipated. She was polite and cordial, even though we basically disagreed about everything. I empathized with a lot of what she was saying, and at a few points along the way, her sense of humor made me laugh. In another space-time continuum, I could picture us being friends.

I still think about that conversation whenever I write about this subject. It made me realize that not all trans activists are malicious and unreasonable. Just as trans activists don't speak for everyone in the trans community, the most vocal and belligerent of these activists don't speak for everyone in the activist community, either. Sometimes it takes a face-to-face conversation to recognize that the other side isn't seeking to destroy all you stand for. In many cases, it only wants to be heard.

Are trans women the same as women who were born women, full stop? The question has undeniably become a land mine with ramifications for sex-segregated spaces like washrooms, locker rooms, and prisons, sports competitions, and our sex lives. Many of the Democratic presidential candidates, including Massachusetts senator Elizabeth Warren, Vermont senator Bernie Sanders, and presidential nominee Joe Biden, were eager to support the controversial Equality Act, which will include gender identity as a protected category under federal anti-discrimination law.

Some have argued that the refusal to acknowledge trans women as women stems from the same kind of closed-minded ignorance that straight women and lesbians once faced in their fight for equality. In activists' minds, it is hypocritical for these women to turn around and hold back the next minority group, now that they have gained equal rights.

The focus of this chapter centers on transgender women more than transgender men because although both groups face a similar issue, the resulting implications of policy are not the same. For example, although trans people across the board endure harassment regardless of whether they are in a washroom corresponding with their birth sex *or* one that corresponds with their gender identity, trans men arguably face a greater threat in men's spaces than trans women do in women's. Similarly, regarding the discussion on sports, the repercussions of including transgender women in women's divisions is not the same as including transgender men in men's.

One dominant narrative is being put forth by each political side and neither is fully correct. Most on the left (not including radical feminists, as we'll see) will say, if you feel a certain way about your gender, that's all that matters; gender is completely detached from biology, so you can identify however you want. It is a continuation of the same antiscientific theme, of gender identity eclipsing biological sex's relevance to human behavior and anatomy. Those on the right will say, your gender identity is entirely tied to your biology and people who are transgender should not be allowed to transition because there is no such thing as identifying as the opposite sex.

What should be the requirement in order for a transgender individual to be allowed in women's spaces? Undergoing a social transition? Legally changing their sex? Is it transphobic that a requirement be met?

Our opinions and policy decisions should be based in reality and facts. Indeed, it feels insensitive to point out that transgender women are, in any way, different from women who were born women, especially when those of us doing the pointing aren't trans. But we can agree that transgender people deserve equal rights and legal protections while acknowledging that some biological differences do exist, and at the same time, these differences shouldn't be used to justify discrimination. Pretending otherwise breeds frustration, not only among those who wish to discriminate against trans people, but those who are supportive of the community, too.

Womanhood

As I mentioned, one of the groups most loudly pushing back against transgender ideology is radical feminists. According to radical feminism, women are a systemically oppressed class and men are socialized into being oppressors by the patriarchy. Gender is a social construct that exists to prevent women from having equal rights, funneling men and women into roles of domination and subordination, respectively. By this logic, transgender women are not, and cannot, be the same as women who were born women because they were born and socialized as male.

If radical feminists believe gender is a social construct, you'd expect they'd also believe that trans women could just be socialized as women after transitioning, but this is not the case. To radfems, the defining feature of womanhood is sex, and the process of transitioning cannot change this; it can only change one's sex characteristics.

On the side of transgender activists, most I've encountered argue in favor of gender being biological, as this supports the argument of having a female brain in a male body and vice versa. Others, surprisingly, will uphold the social constructionist argument, saying that trans women have learned what femininity is through exposure to gender stereotypes dictating what it means to be a woman.

Regardless of whether a person believes gender is biological or socially constructed, the discussion revolves around whether a person's gender is necessarily in alignment with their anatomy. As we'll see, radical feminists and conservatives, who would otherwise have next to nothing in common, have found common ground when it comes to this issue, as both believe gender is something that can't be changed in reference to biological sex.

It is a bitter battle, with threats of violence championed in slogans like "Punch a TERF," and online arguments deteriorating into women threatening to cut off trans women's genitals if they should step into women-only spaces. The fact that this is taking place on the left side of the political spectrum—both the transgender community's hostility and aggression toward feminists, and feminists' retorts that the word "TERF" is a slur and trans women are in fact misogynistic, entitled men—has led many on the sidelines to sit back, munching on popcorn, while musing about how progressives are content to eat their own.

Though it may seem like an obscure discussion, this leftist infighting highlights an unspoken hypocrisy. A political movement that is supposed to be championing the rights and freedoms of women has instead been steadfast in accepting their subjugation and erasure. Radical feminists are resentful that women and girls are being treated like second-class citizens when it comes to physical safety in their own spaces, and womanhood is actively being redefined by those whom they consider to have male privilege. Not only that, but to radfems, trans women embody and promote the most regressive stereotypes of women being hyperfeminine, deferential to men, and scantily clad, and trans women have not had to endure what it really means to be a woman: sexism, sexual harassment, and having to live in fear of men.

Another example of erasing women's identities arrives in the form of gender-inclusive language. Gender-inclusive language avoids any reference to the words "woman" or "female" or female body parts, since trans women do not necessarily share this anatomy. Transgender activists and their allies will use the word "cissexism" (stemming from the word "cisgender," which means not transgender) to describe this as a form of discrimination, one that operates from the baseline assumption that everyone identifies as their birth sex. Cissexism is similar to transphobia, but is seen as subtler.

It is considered transphobic to say that women have vaginas, give birth, and have periods. Instead, a new set of language has been devised, including dystopian-sounding terms like "pregnant people," "birthing parents," "uterus-bearers," and "menstruators." (Much outrage ensued after the word "front hole" appeared on a health information website, but it turns out its use was only meant to offer an alternative, and not a replacement, for "vagina.")[1] Women are being told not to celebrate motherhood because it can be alienating to trans women.

Radical feminists, however, are not having it, calling it an infringement on women's liberation. Using female language to describe women's experiences is regarded as empowering and validating, especially around matters pertaining to reproductive rights. For example, issues like abortion relate only to women who were born women, and not transgender women. You might be thinking, "What about trans men?" since some can and do become pregnant. Radical feminism is open to trans men because by its definition, trans men are female. Trans men will, however, be referred to as female, not male.

Radical feminists have been at times inflammatory in their refusal to recognize transgender people as the opposite sex. They will refer to trans people by their birth name (also known as "dead-naming") and use the pronouns associated with their birth sex (for example, calling a trans woman "he"). Instead of referring to someone who is transitioning to female as a trans woman or "male-to-female," they will use terms like "MtT," which stands for "male-to-trans," or "TIM," which stands for "trans-identified male." (Similarly, trans men are called FtT and TIF.) Radical feminists also won't use the word "cisgender" when describing people who are not transgender.

I don't believe transgender women should be reduced to their biology or be referred to as male. At the same time, it isn't accurate to say that transgender women are biologically women, as some trans activists claim. Neglecting these differences is unhelpful, especially for trans women, considering that they are predisposed to particular health conditions based on these biological differences that natal women are not. In order to offer the best care possible, medical professionals need to obtain an accurate patient history.

William Malone, the endocrinologist I interviewed in the last chapter, told me, "More than five percent of all records of transgender persons presenting for inpatient hospital care have inconsistent sex markers." In one widely publicized case, a transgender man arrived at the ER, presenting with abdominal pain. The ER staff didn't realize that the individual was transgender or that his pain was due to being pregnant. Due to this delay, he lost the baby.

Another example of why these differences need to be acknowledged is prostate cancer. Besides skin cancer, prostate cancer is the most common cancer in American men; one in nine will be diagnosed in their lifetime.[2] Transgender women, even after transitioning, face a higher risk of prostate cancer than women. In one case study, a transgender woman who had had her testes removed was diagnosed with prostate cancer forty-one years after her transition.[3]

Saying that trans women's body parts are biologically female obfuscates the reality of such a risk and increases the likelihood that important medical practices associated with biological sex will be overlooked. Similarly, it should be perfectly acceptable to say that some men have a uterus, so long as it doesn't preclude an honest discussion as to why (that is, he was born female).

Are Sexual Preferences Transphobic?

Is having a sexual preference inherently transphobic? One of the most heated areas of this debate pertains to how this new language plays out in the bedroom. (No, not in the fun way you're thinking.) This conversation applies mostly to lesbians and straight men, as transgender women have been more outspoken about their difficulties in attaining sexual partners than transgender men.

It is certainly trendy nowadays for gay people to refer to themselves as "same-gender attracted" instead of being same-*sex* attracted in order to be more inclusive of trans people. More specifically, if someone identifies as a same-*gender* attracted gay man, it means he's attracted to everyone who identifies as a man, regardless of whether they were born male.

It's a minor change in terminology but it alters the meaning significantly. I am in complete support of people dating and sleeping with whomever they like. (I will always be sex-positive, first and foremost.) I do think that transgender people face greater prejudice in the dating world, and we should all be on board with wanting to change this. The issue I have is that sexual orientation is about sex, not gender, and *no one* should be made to feel as though their consensual sexual preferences aren't legitimate.

The first time I heard of tensions between transgender women and the lesbian community was over something called "the cotton ceiling." The cotton ceiling is an allusion to the glass ceiling that women face, but in this case, it's referring to underwear. Transgender women who are sexually attracted to women are also considered lesbians post-transition, and many have a difficult time finding women who will sleep with them if they have not undergone gender reassignment surgery.

This is because most lesbians, by definition, are not sexually attracted to people who have a penis. Penile-vaginal intercourse is technically heterosexual sex, which conflicts with the definition of being lesbian. Attempts to coerce lesbian women into being sexually interested in penises have been rightfully compared to conversion therapy and corrective rape.

For having sexual boundaries, lesbian women have been called transphobic and—just when you thought things couldn't get more inane—vagina fetishists. But if genitals don't matter, as some of these trans women claim, then why aren't they having sex with each other?

In response, organizations like Get the L Out and the LGB Alliance have mobilized in the United Kingdom, emphasizing that gay people are same-*sex* attracted and that they, along with bisexual people, have little in common with the transgender and queer factions of the "LGBTQ" initialism.

I understand trans women's frustration. It's not fun being rejected in the dating marketplace. A study from 2018 showed that transgender people are indeed often excluded from the pool. In a sample of more than nine hundred people, including gay, straight, bisexual, and trans individuals, close to 90 percent of respondents stated they would not consider dating a transgender person.[4] Nontransgender straight men and women were the least likely to consider it.

How do we balance our personal preferences against wanting to lessen this discrimination? In my opinion, what's important is that someone who may be open to dating or having a romantic relationship with someone who is transgender isn't dissuaded from doing so out of intolerance or narrow-mindedness. In my time as a sex researcher, countless men I interviewed would tell me that they had had trans women as partners in the past and felt shameful about it. They didn't feel they could be open about it or introduce their partners to their friends or family. Needless to say, it isn't right to treat a partner differently because they are transgender, nor is it acceptable to hide your girlfriend or boyfriend from the people in your life because you're afraid of how they'll react.

From a scientific perspective, those who are sexually interested in women with a penis have a paraphilia known as *gynandromorphophilia*.

By saying this, I don't want to endorse or encourage the fetishization of transgender women. Trans women, in particular, face tropes of being hypersexual and of "tricking" straight men into having sex with them. (Asian women are similarly fetishized by the media and society. I am not transgender, so I can't speak to the experience of being fetishized for being trans, but I can understand where the community's concerns are coming from.)

It isn't accurate to claim that women and men who are attracted to women should be interested in dating a woman with a penis. At the risk of sounding like a broken record, the expression of our sexual preferences is culturally influenced, but our preferences are, beneath it all, biological. Some might argue that attraction is a based on a number of factors beyond vaginas and penises, including personality, common interests, and chemistry, and that we may be open to dating someone who has physical attributes, like hair and eye color, that don't necessarily fit our usual type. I would agree—yours truly used to only date scary-looking guys covered in tattoos and facial piercings.

But all of these characteristics are subsumed within an individual's sexual preference for men, women, or both. In many cases, practical elements, like wanting to start a family, can also influence one's choice of partner. It is not due to nebulous, ill-defined concepts like the "heteropatriarchy," the idea that straight men are allowed to dominate in society.

For some people, genitals and anatomy don't matter as much, and they are open to dating people regardless of their sex *or* gender. But this isn't the case for everyone, and it isn't fair to impose that expectation on a potential partner. It shouldn't be considered bigoted to have sexual preferences. Gender is only one part of a person's identity, whether a person is transgender or otherwise, but when it comes to our sex lives, this isn't a negligible factor, nor should it be. Browbeating people about their sexual preferences does little to win them over to the cause of equality.

These are the harsh realities of dating. Someone might reject you for frivolous reasons relating to your physical appearance, like a face mole or crooked teeth. No one should have to justify the choices they make about their bodies, nor the people they will and won't be intimate with.

Our sexuality should not be policed on either side, and I have heard from some transgender women that they have no issue finding lesbians who are interested in dating them. It only becomes a problem when other people—including, in some cases, radical feminists—start interfering, threatening to revoke the partner's status as a lesbian if she is willing to have sex with a woman with a penis.

Washrooms, Prisons, and Shelters

Following from the question of what makes a woman, should trans women be allowed in women's spaces? In 2016, North Carolina was the first state to pass a bathroom bill law, House Bill 2, that barred transgender people from using their preferred restrooms if they had not undertaken a legal and surgical change. In 2019, a federal court overturned this law following public outcry and economic backlash for the state.

The reversal is symbolic of the wider trend regarding where policy about trans-inclusive spaces is headed and whether transgender people should be allowed to access sex-segregated facilities based on their gender identity instead of their biological sex. Those in opposition to these policies raise concerns about the likelihood that nontransgender men will exploit the law, posing as trans women, in order to sexually victimize women and girls.

The knee-jerk reaction from most progressives has been to dismiss these fears as overblown—only the mind of a coldhearted bigot would think such a thing and cast aspersions on policies designed to make life easier for trans people.

In truth, gender-neutral spaces *are* more dangerous for women and girls because they pose an opportunity for sexual offenders to gain access to victims. Recent statistics show that almost 90 percent of complaints about sexual assault, voyeurism, and harassment in United Kingdom swimming pools happen in unisex changing rooms.[5] (The word "unisex" has also become forbidden because it references sex instead of supporting gender as a distinct entity.) There have also been numerous instances of sexual predators taking advantage of gender-neutral restrooms and changing rooms in order to spy on, record, and expose themselves to unsuspecting girls and women.[6]

In 2012, a convicted sex offender managed to gain access to two shelters in Toronto by claiming to be a transgender woman. After pleading guilty to sexually assaulting two women, he was given an indefinite prison sentence and a dangerous offender designation. His previous crimes included sexually abusing a five-year-old girl and raping an intellectually challenged woman.

It's important to note that the main threat to women in gender-neutral spaces does not come from transgender people, but from *men* invading these spaces. As someone who has worked with violent sexual offenders in a research and clinical capacity, a hallmark trait of men who commit sex crimes is antisociality, a blatant lack of remorse about hurting others.

We should also take into account that transgender women have endured an unfortunate history of prejudice and being unfairly painted as sexually deviant. Most important, transgender people should not be penalized when antisocial men choose to manipulate the system. Even in the event that someone experiences autogynephilia (discussed in Chapter 4), I don't believe they should be prohibited from accessing women's spaces, because autogynephilia has nothing to do with sexual coercion. It is antisociality, not the existence of a paraphilia, that leads someone to commit a sexual offense. (This is even true in the case of a paraphilia like sexual sadism, wherein individuals gain sexual gratification from hurting their partners.)

The harassment and abuse indeed goes both ways. In addition to exacerbating feelings of gender dysphoria, being forced to enter men's spaces can put transgender women in danger, seeing as how some men may not take kindly to them being in the vicinity. The same can be said for transgender men if they are forced to use women's facilities.

But instead of having a nuanced discussion so that sensible policy can prevail, the entire conversation has been brushed aside as irrelevant, leading to unnecessary divisiveness between political camps and animosity toward both the trans community and opponents of these policies.

It doesn't seem like we will be having a reasonable discussion anytime soon. We have some folks like Joe Biden saying that sex is based on how you identify, not what prison says. During the LGBTQ Presidential Forum in 2019, Biden remarked, "In prison, the determination should be that your sexual identity is defined by what you say it is, not what in fact the prison says it is." This may not be the most enlightened move, as we saw with one convicted rapist and child molester in the United Kingdom, who was placed in a women's prison and subsequently sexually assaulted four female inmates.

In the United Kingdom, police forces now record a male-born rapist as female if the individual wishes to identify as a woman.[7] In the county of Kent, police reported managing three offenders convicted of rape who self-identified as female. I imagine this number will increase sharply in the prison population in the coming years. Already, 1 in 50 male prisoners in the British prison system identifies as transgender.

Don't get me wrong—I am sympathetic to the risks that transgender women face being placed in men's prisons. Inmates are not typically known for their kindness and hospitality. Consider, however, that a sixteen-year-old school shooting suspect in Colorado, who was born female and had transitioned to male, was housed in a *female* detention center and likely won't be sent to a male prison if convicted, highlighting the double standard in this entire debate.

Although it can seem unsympathetic to require someone to undergo a physical or legal change in order to "prove" they really are who they say they are, policies across the board need to be based in more than self-identification. It is an imperfect answer because in some cases, a person with gender dysphoria may not be able to afford medical interventions. There is also a chance that a perpetrator who is *not* gender dysphoric will take these steps in anticipation of being charged with an offense. It doesn't make sense to stratify male prisoners who have been previously convicted of a sexual offense, because many who do commit such offenses aren't necessarily caught.

The best solution would be opening prisons or units dedicated to transgender inmates, as was recently done in London.[8] There will still exist a pecking order regarding risk of being assaulted, but that is, to some extent, unavoidable, even among nontransgender populations in prison. In addition, mental health professionals should be supported in doing thorough and accurate assessments to determine whether an inmate is truly gender dysphoric or malingering, in addition to assessing for other mental health issues, like antisociality and, perhaps most important for those convicted of sexual offenses, their risk of reoffending. In the event that a transgender prison isn't available, decisions should be made on a case-by-case basis regarding whether an inmate is more appropriately housed with women or men. Sweeping policy decisions in either direction are unhelpful.

As it stands, women and girls continue to be punished for holding an opposing view, leaving them little recourse. It is a debate that has pitted trans activists and their allies against both social conservatives *and* left-leaning radical feminists, who have joined alliances to support bathroom bills and argue that female-born women and transgender women are not the same. I could barely believe my eyes when I saw that the conservative organization the Heritage Foundation hosted the radical feminist group Women's Liberation Front in a 2019 panel.

In another such example from 2018, a Toronto woman, Kristi Hanna, was forced to leave a women's shelter after being assigned to share a room with a preoperative transgender woman. At the time, Hanna inquired about her legal rights with the province's human rights support center; as someone with a history of sexual trauma and substance dependence, she found the arrangement detrimental to her well-being. She was told, by referring to her roommate as male, that she was the one engaging in discrimination. Hanna was given the option to stay in a different room that led to a fire escape and didn't have a door that closed. As a result, she left the shelter.[9]

In a situation where there is no middle ground, who should be expected to give? The question can be distilled down to whether jeopardizing the safety of one group is an acceptable risk in the name of forwarding the rights of another. How do we enable both women and transgender people to live their lives as they please?

With the rise of the gender-neutral or "all-gender" washrooms, one might suggest offering a separate facility so that laws don't require altering and the whole complicated issue of safety can be avoided altogether. Nowadays, it isn't sufficient to offer a third space because this isn't considered true inclusivity. Activists aren't satisfied with being treated as separate but equal—they want everyone to be treated exactly the same.

The discomfort is much more common than it seems. I have had women tell me quietly that they do not want public washrooms to become gender-neutral because they become less safe, comfortable, and clean. I personally have noticed restaurants and other buildings in my hometown of Toronto making the shift to all-gender washrooms, which usually consist of a row of stalls with a communal sink. The door of the stall will have a lock, but the door doesn't always reach the ceiling and the ground.

Some might say, "Everyone has an all-gender bathroom in their house," to which I say, *Really?* They're going to compare a single-occupancy bathroom in someone's house to restrooms in a public space? Even if multiple people are sharing a bathroom in a house or an apartment, you all know each other, or at least know that anyone using the bathroom knows someone you're living with.

Ultimately, I believe in live and let live. If a transgender individual is in the presence of someone who appears to be uncomfortable, whether it's in a bathroom, locker room, or shelter, let it be up to the trans person to decide what's appropriate.

We should be compassionate, but not to the point of overriding common sense. We must allow the conversation to exist. As it stands, any questioning of progressive policy and bringing up concerns about safety is immediately labeled as hateful. It isn't fair to throw women and girls under the bus, telling them their concerns don't matter.

Sport

Though many will grit their teeth and keep their mouths shut about various overzealous instances of transgender activism, sport has been one area where the fountain of civility has run dry. Some of the most progressive people I know, who have remained silent about transitioning kids, gender-free bathrooms, and terminology like "pregnant people," will outright say it's unfair for trans women to be competing in women's sports.

A transgender woman, Veronica Ivy (previously Rachel McKinnon), recently defended a world championship title in women's cycling after breaking two world records, sparking an international debate. We see trans women annihilating their competition and shattering records in every sport fathomable, from power lifting to handball, volleyball, track, and one of my favorite sports, mixed martial arts. From my observation, every time it's been revealed that a transgender woman is competing in the women's category, fury ensues, with fans of the sport condemning the decision and mainstream news outlets attempting to reassure the public that there is nothing unfair about it. There have been no easy answers to date, and in some cases, the dispute extends beyond the competition itself. In 2019, transgender power lifter JayCee Cooper sued USA Powerlifting for discrimination after it barred her from competing.

In the eyes of everyone watching, sport is supposed to be a meritocracy and a level playing field. Professional sports are divided by sex for a reason, especially at the elite level. Without sex divisions, women would not have a fair chance at winning against men.

To question whether these new allowances are fair has been decried as hate speech. Female competitors will denounce transgender women competing against them, but few will go on the record saying so, for predictable reasons—for example, tennis champion Martina Navratilova was taken to task on social media for criticizing the allowance of transgender women who have not undergone surgery to compete as women.

I strongly believe that transgender people deserve the same opportunities as everyone else, including in the realm of professional sports. But it shouldn't be deemed off-limits to question whether someone who identifies as the opposite sex still has some physical characteristics in alignment with their birth sex, since biological sex can't be changed, even for someone who has transitioned.

The International Olympic Committee (IOC) guidelines are considered the gold standard that all other athletic organizations follow. Its current transgender guidelines were issued in 2015, stating that trans women must get their testosterone levels below 10 nmol/L for twelve months prior to competing as women and that no surgery is required.

Elite female athletes tend to have levels below 1.68 nmol/L. For anyone who wants to argue that "women have testosterone, too," trans female athletes are being permitted to compete with testosterone levels almost six times higher than what would be found in their fellow competitors. Not only that, but a new study from Sweden's Karolinska Institute found that suppressing testosterone in trans women does not reduce muscle strength, even after a year.

As it stands, the current iteration of guidelines is still being debated because the IOC's scientific committee has been unable to come to a consensus due to the politically charged and highly emotive nature of the issue. Some have argued for a lowering of this limit from 10 nmol/L to 5 nmol/L.[10]

This is no trivial matter, considering that, for many high-contact sports, competitors risk being badly hurt, and even killed, if they are overpowered by their opponent. According to the Association of Boxing Commissions, transgender athletes who have undergone male puberty can fight in the women's division if they have had a gonadectomy (removal of the testes) and have been taking hormones for two years post-surgery.

Most of these guidelines do not take into account the difference between organizational and activational hormones. Organizational hormones affect our development, leading to changes that are irreversible. Activational hormones circulate throughout our body, and their effects are mostly reversible. In the debate around transgender women in sport, activational hormones are being treated as though they are the only kind that matters.

Identifying as female doesn't negate the advantages an individual has gained from undergoing male puberty, including those related to greater height, upper body strength, wrist size, hand size, muscle mass, lung capacity, and bone density. Males are, on average, stronger, faster, and larger than females.

Testosterone increases muscle mass, leading to greater strength and endurance. Putting estrogen in a person's body—even if they have removed their testes—doesn't override the benefits of having this structural foundation. It also doesn't mean that someone will perform at the same level as women. It is estrogen—not testosterone—that causes bone growth, which means that even if a transgender woman has lowered her levels of testosterone and has had gender reassignment surgery, undergoing estrogen therapy can compensate for her presumed loss in bone mass.

Those supporting the classification of transgender women as female athletes are essentially asking all of us to ignore this information in the name of granting acceptance. You would need to be clueless about biology or ideologically possessed to not see why this is unfair.

I understand the desire to be empathetic toward transgender athletes. The goal of treating them no differently from athletes who are not transgender is admirable, and using physical differences to inform which category they should compete in feels callous, as though we are telling them which gender they are. But using emotion to drive these decisions is not rational, reasonable, or harmless.

Some have shrugged off these advantages as insignificant, saying there is much overlap between the sexes, seeing as how some women just happen to be taller or naturally stronger than men. Others classify the competitive edge that a transgender woman has as a normal variation you'd find among athletes who tend to be gifted with unusual biological advantages, like swimmers with large feet or basketball players who are extremely tall.

We all know of women who are taller and stronger than some men, but these women are outliers. They don't change the fact that the average man is taller and more physically powerful than the average woman. A perfect illustration of this would be the tennis match between Karsten Braasch, a former professional tennis player who was born male and played in the men's league, and Venus and Serena Williams at the 1998 Australian Open. The Williams sisters had announced they could beat any male player ranked in the Top 200. Braasch, who ranked 203rd at the time, took on the challenge and beat them both without difficulty.

For transgender women, some may not have a physical advantage over another female competitor, as we saw in the case of Ashley Evans-Smith, an American mixed martial artist who was born female and fought in the women's division. In 2013, Evans-Smith beat transgender female fighter Fallon Fox. But it is far more likely that the average trans woman will excel over the average nontransgender woman. The advantages that trans women have from being born male are not just flukes bestowed from nature; they are because individuals born male belong to a different physical category than those born female.

As for intersex athletes, they should be given consideration in a manner separate from those who are transgender, regardless of whether an intersex individual also identifies as trans. Intersex athletes experience differences in their anatomy and testosterone levels due to their biology in a way that transgender athletes do not.

Should intersex athletes be admitted into women's competitions? It depends on the particular condition. Having high levels of naturally occurring testosterone can enhance one's performance, and for someone with androgen insensitivity syndrome (AIS), they may have an advantage, depending on whether they have complete AIS or partial AIS.

Someone who has complete AIS has XY chromosomes, typical of males, but their androgen receptors do not function as expected. As a result, their body does not respond to testosterone during development. They have an outward appearance of being female and testes that are either undescended or partially descended. An individual with partial AIS, on the other hand, has XY chromosomes, and their body partially responds to the testosterone. (The reason for this distinction will become clearer in a second.)

According to guidelines published by the International Association of Athletics Federations, which seek to level off any advantages an intersex athlete may have, someone with partial AIS will be affected only if her body responds to the elevated testosterone. If so, she will be required to lower it, say, by taking hormonal birth control. An athlete with complete AIS will not be bound by these restrictions, because she does not benefit from the testosterone in her body and won't have an advantage over other women.

It's extremely unfortunate that as part of this ongoing debate, a number of intersex athletes have had to withstand public scrutiny and speculation after their personal medical information was leaked. The decision-making process and private medical records aren't anyone's business. Going forward, we should hope that every decision will be made privately and confidentially between an athlete's team and the governing body.

The cutoffs of these guidelines go beyond partisan politics and have real consequences for athletes' lives. For girls who are competing for college scholarships, who have been training their entire lives for these opportunities, it is particularly demoralizing to be competing against transgender girls, knowing that the cards are stacked against them and the outcome is predictable even before the competition has begun.

As a final point, I want to emphasize that I do not support mockery of transgender or intersex athletes, or discriminatory comments made about them based on physical appearance. The rhetoric is unnecessary and doesn't add anything of value to an argument that already has merit in its own right.

I also don't believe that most trans athletes are transitioning with the intention of having an unfair advantage. Nevertheless, it is women who will lose out more than men with these rules because transgender men do not pose a threat to men's sports. As these policies become more widespread, many foresee the end of women's competitions.

Yet again, science denial is perfectly acceptable if it follows in the fashion of woke politics. At a time when society is actively pushing for the equality of women, this entire debacle is flabbergasting, highlighting exactly who is calling the shots in the identity politics hierarchy. There is nothing wrong with advocating for meaningful and fair opportunities for *everyone.* They should not come at the cost of prioritizing one group over another.

THE FUTURE

MYTH #7

WOMEN SHOULD BEHAVE LIKE MEN IN SEX AND DATING

As an outspoken sexpert, people frequently ask me for my advice about sex, dating, and love. This tends to happen for two main reasons: 1) my academic background is in the science of sex, and 2) they know I'll be brutally honest.

Sex and relationships are no easy task, especially today, considering the collection of countless dating apps we all have on our phones. The process of getting to know someone can be messy, clumsy, and confusing. If that wasn't challenging enough, the dating landscape is rapidly changing. For young people (and particularly the 97 percent of us who are straight), they've been stranded on an island of misinformation, inundated with unhelpful ideas about what they and the other sex want.

The Denial of Evolution

The modern-day feminist movement maintains that a well-functioning society should do away with gendered norms. In theory, this would be a good thing; a person should be free to live as they please without being held to expectations based on an immutable characteristic like sex.

But then this line of thinking goes too far—if the sexes are truly equal, it reasons, there should be no differences between men and women, including in the realms of dating and sex. Any deviation from this symmetry, or any adherence to gender-typical roles, is seen as chauvinistic or "toxic" of men and belittling of women. The admirable goal of gender equality has metastasized into meaning that men and women are the same, and therefore, any sex differences are socially imposed and can be unlearned, or altogether don't exist.

As a result, it's become fashionable to call into question biology and evolutionary psychology, which is a shame because they offer a truly enlightening perspective into the way we think. Both have become verboten in recent years, tarnished with allegations of sexism, irrelevance, and being in opposition to women's rights. In actuality, nothing could be further from the truth.

The sexes have different sexual systems, stemming from thousands of generations of evolutionary influence, and these differences play out in meaningful ways. Evolutionary explanations may be used by *some* individuals to disrespect women and deny female autonomy, but once again, instead of pretending that these facts are falsehoods, a more productive approach would be to call out people who use them to promote backward ideas.

It's totally fine to be critical of social norms—and in fact, I think we should all be skeptical, to some degree—but it's harmful to replace truth with wishful thinking in the name of a larger political goal. The denial of evolution is one of the most damaging myths I've encountered, because it throws everything we know about human sexual behavior out the window. It actually *hurts* women to behave like men, and vice versa, when it comes to courtship and sex. This is the case regardless of one's political affiliation.

For the record, I am certainly not a traditionalist. In addition to my research expertise as an academic scientist being on the subject of kinky sex, much of my writing advocates for sex positivity and comprehensive (as opposed to abstinence-only) sex education. I believe the focus of sex should be pleasure and that consensual sex without intimacy or commitment is perfectly acceptable. On a personal level, I've always been career-driven, more focused on financial independence than finding a husband.

But for some reason, drawing upon biological or evolutionary explanations for behavior, or even hinting that abiding by traditional gender roles might actually *help* us when it comes to romantic relationships, leads to allegations that I also think a woman's only purpose in life is to look pretty, have babies, and be subservient to men. I'm not suggesting any of these things, and anyone who knows me would laugh at how preposterous this assumption is. I'm highlighting what the data are telling us because single, young people are experiencing mass confusion.

What Makes an Attractive Partner?

The evidence supporting evolutionary psychology is so overwhelming, it's impossible to dismiss it. In an ideal world, this research would be innocuous and uncontroversial.

The act of sex comes with a greater cost to women, due to the possibility of becoming pregnant and having to take on related responsibilities—giving birth, breastfeeding, raising the child, and ensuring his or her survival. By comparison, for men, sex requires an investment of several minutes. (About five or six minutes on average, according to recent studies.)

As a result, the female sexual system evolved to account for this discrepancy, because women who made good mating choices were the ones who succeeded at passing on their genes. This is why women, on average, are more selective about their sexual partners, preferring those who possess status and resources that will benefit them and their future offspring. Back in the day, if these resources were of poor quality or inconsistent, this would threaten a woman's survival and the survival of her children. Being less choosy about sexual partners could also result in raising a child without the help of the father, who could otherwise provide material resources in addition to emotional support and physical protection. In response, men have evolved to be highly competitive in order to be attractive to the choosiest (and, therefore, highest value) sexual partners.

Charles Darwin's theory of natural selection describes the process by which individuals who are well adapted to their environment are more likely to survive and reproduce than those who are less well adapted. Sexual selection, one of Darwin's critical revelations, maintains its relevance, stating that the mate preferences of one sex will determine the characteristics that are passed on in the other sex. For example, our female ancestors preferred male partners who could offer physical protection to them and their young, which is why greater height is generally seen as an attractive trait in men.

Also relevant to the discussion is Robert Trivers's "parental investment theory."[1] Eggs are more costly to produce than sperm; consequently, women release one egg a month while men produce several hundred million sperm a day. It then follows that 1) women are very selective about who gets to fertilize that egg, and 2) men need to be competitive in order to get access to this precious resource.

Also, because women's reproductive rate (the amount of time it takes to produce offspring) is slower than men's (nine months versus five or six minutes, as noted earlier), this amplifies the competition among men for female partners. Since women are the arbiters of sex, from an evolutionary standpoint it would benefit men to be drawn toward pursuing sex with multiple partners.

Before I go any further, it's important to note, like any other scientific finding, statistical trends don't speak for everyone. Individual differences abound, and when it comes to our sex lives, there are some women who behave more like men, some men who behave more like women, and some people who don't feel like they fit into either line of behavior (although in this case, I'd argue they likely aren't being truthful with their partners or themselves). More important, regardless of whether a particular behavior has been useful in the past for the propagation of our species, this doesn't give anyone permission to be a jerk in their relationships.

Although the advent of birth control in the 1950s has allowed women greater sexual freedom—and in some cases, men will become the primary caregivers, and in other cases, women will choose not to have children at all—we have not been able to override millions of years of evolutionary influence on our mating strategies. We may be subconsciously driven toward particular traits or tendencies, but how we treat our partners is a conscious choice, and individuals should be held accountable for their behavior.

Some scholars will argue that these preferences are due to socialization because a patriarchal society has left women more economically disadvantaged than men, and historically, women have been dependent on them. Others accuse evolutionary theory of being just that—a lightweight "theory" that shouldn't be given much thought.

This is an example of a word meaning different things to different people. To most nonscientists, a "theory" is just a random idea that's been strung together, that may or may not have any basis in reality. In the scientific community, however, a "theory" is an explanation that has been tested with experiments to determine whether it is true. Scientific theories are never considered proven, but rather, continue to stand until they are *disproven*.

As for the belief that sex differences in mate preferences are socialized, cross-cultural research would beg to differ. A study sampling more than 10,000 women and men in thirty-seven countries across six continents and five islands found that women placed more importance than men on a partner's financial prospects. This was the case regardless of where in the world these women lived or under which political system, including communism and socialism.[2] For women who are successful, they place an even greater emphasis on their partner's generosity and financial success. This explains why some men will purchase expensive toys, like flashy watches and cars, as a way of signaling status to potential partners.

I remember having a discussion with two male friends when I was in graduate school about whether a man should pay on a first date. I argued that he should, because this signals to a woman that he is serious about her. A woman usually looks for signs of commitment and that a man has the means to provide for her.

Both of my friends were aghast, decrying this as sexist, saying they would only pay for a date's drink or meal once they knew her and knew that they were interested in dating. Despite the fact that both wanted serious relationships, neither they nor any of their peers apparently subscribed to my way of thinking. They asked me quizzically if I had been dating fifty-year-olds.

With the mainstreaming of feminism, it seems millennial men have also bought the lie that gender equality requires the sexes be treated identically. What does this mean for women? I would argue dismay and disappointment, which I will elaborate on shortly. As for my friends' comments, I should add, there's nothing wrong with intergenerational relationships. In fact, research shows that most women prefer to date men who are older than them because a man's age is correlated with his financial security.

Feeding into this preference is hypergamy—women have a tendency toward trading up in the hopes of getting the best partner they can find. Women desire more than just sex in a committed relationship; they want their emotional and practical needs to also be met. As a result, when women cheat, it is usually due to a lack of emotional intimacy in their current relationship or because their sights have shifted to someone who is more successful than their current partner. It's about more than lust or the physical act. When men cheat, however, it's for sexual novelty—the desire to have sex with someone new.[3]

Now, I am definitely not saying this excuses infidelity, that cheating is inevitable, that men don't have emotional needs, or that these are the only reasons people stray from their partners. Regardless of what evolutionary trends predict, deciding to have sex outside of a committed relationship is unethical. A more honest approach would be to talk to your partner if you are feeling unhappy in your relationship, or to end things altogether before getting together with someone else.

As for men's sexual preferences, they are a solution to an interesting evolutionary quandary. Because men are unable to tell, simply by looking at a woman, whether she is of high fertility (able to produce healthy offspring) and high reproductive value (able to have many children), they've evolved to prefer partners possessing observable qualities indicative of both.

Now, I don't believe a woman's only value is her youth or fertility, and I do believe that cultural norms should be willing to change. But the only way this can happen is by acknowledging where the drive is coming from. Contrary to what "woke" advertising and media tell us, beauty standards aren't socially constructed, arbitrary, or due to living in a "heteronormative society." Men tend to prefer partners who are young and beautiful—possessing symmetry, smooth, clear skin, and shiny hair—because youth and attractiveness are physical cues that a woman is healthy and will have greater reproductive success.

Explanations centering on socialization claim that men prefer younger partners because they are easier to manipulate and control. This may be the case for some men, but on the whole, science would suggest otherwise. When we look at male teenagers, if manipulation and control were the driving forces in mate choice, they should prefer female partners who are younger than them, who would be more naïve and less experienced. Instead, young men report being more sexually attracted to women who are slightly older, who would, in evolutionary terms, be more reproductively fit than younger girls.

In the name of being thorough, I'll also explain *gerontophilia*, a sexual preference for older adults, age sixty and up. In rare cases, some men will prefer elderly partners, but this is a paraphilia (like autogynephilia and gynandromorphophilia, mentioned in previous chapters). Like many paraphilias (such as a sexual interest in latex or feet), gerontophilia is unusual because being primarily sexually interested in someone who is not of childbearing age is not evolutionarily advantageous.

I don't dismiss the validity of concerns about male predation and sexual coercion, as they are real problems in society. I have encountered men who strictly date eighteen-year-olds (hovering as close as possible to the lower boundary of age of consent laws) whereas they themselves turned eighteen a long time ago. Most emotionally healthy, well-adjusted men will prefer a partner who is closer to them in age so that they will share a similar level of maturity and, quite frankly, be able to have a conversation.

This explains, in part, why the beauty industry is today a $500 billion industry. Full lips and high cheekbones are a sign of facial femininity, linked with higher levels of estrogen and fertility. This may also explain the growing popularity of lip injections and women overdrawing their lips on social media. Contour and highlighting exaggerate facial contrast to make one's features appear more feminine, and foundation and concealer hide under-eye circles and discoloration to give the appearance of an even complexion. This is not due to the patriarchy or capitalism trying to sell women things they don't need. These are tendencies that are, to some extent, hardwired in us.[4]

Another common indicator of attractiveness is waist-to-hip ratio (WHR), because it offers clues to a woman's reproductive history and reproductive potential. Cross-cultural research has shown that, no matter where you are in the world, a low WHR (evident when a woman possesses a small waist and larger breasts and hips) is found to be attractive. The most attractive WHR, according to studies, is about 0.70.[5] WHR tends to be lower in women who have fewer children, and also, women who are not currently pregnant, offering a visual estimate to an interested suitor regarding the likelihood that a woman would conceive and bear healthy children if he were to have sex with her. It is very unlikely that men have been socialized on a global scale to find an hourglass-shaped figure attractive.

In modern-day beauty trends, this may explain the appeal of having a big "booty," as popularized by celebrities and social media, and the subsequent rise in Brazilian butt lift cosmetic procedures, in which fat is taken from other parts of the body to create a more voluptuous bottom.

The field of neuroscience has similarly offered evidence for these preferences. In one study, the nucleus accumbens, a brain region associated with feelings of pleasure, was shown to activate when men looked at attractive female faces.[6] Research has also shown on-average sex differences in brain activation during sexual arousal. When men and women are brought into the laboratory and are shown pornography during an fMRI scan, visual areas of the brain are more activated in men, and regions associated with inhibition and emotional processing are more activated in women.

From an evolutionary standpoint, these differences would be expected because they serve us quite well. At the same time, this doesn't justify bad behavior, like being disrespectful toward one's partner, leering or rubbernecking at other women, or promulgating stereotypes that women are "hysterical," "frigid," or too emotional about sex.

I also think it's a bit unfair that women get more flak from society for their partner preferences than men. Women who date wealthy men are commonly derided as gold diggers, while men who prefer to date young, attractive women will get an eye roll at most. The response from those seeking gender equality should not be to tell women they need to reorganize their priorities or to shame men for being heterosexual, but to push back and ask why women should be held to a different standard or be punished for what they value.

Overall, for both men and women, a partner's looks matter because having a child with someone who is attractive means your children will inherit those good looks and be in a position to attract better mates in the future, thereby ensuring that your lineage continues.

At the same time, the greater emphasis placed on women's looks can feel unjust, because—let's be real—it requires a lot of work. Men can argue that they, too, have to worry about shaving and perhaps some form of bodily hair removal in addition to getting dressed and styling the hair on their head. But it becomes a form of advanced sport when you are a gender-typical woman, encompassing all of the above in addition to makeup, accessories, manicures, high heels, and fifteen different products in the name of skin care.

I often joke about these differences with my male colleagues, who can breeze through airport security in under a minute when it takes me twice as long to simply unload all of the travel-sized containers from my carry-on. One time, I was scheduled to do a remote TV interview and the hairstylist was running late. I had to go on camera with my curly, bleached mess of hair looking like it had been chewed on by a small animal. For the next week, people who had seen the show were understandably concerned, asking if I felt all right and when I had plans to see a hairdresser.

From a serious perspective, though, the increased pressure on girls and women to maintain their appearance can have harmful consequences. Studies of adolescent girls have shown increased rates of disordered eating, negative self-image, depression, anxiety, and poor self-esteem associated with prolonged media exposure. Over time, the extra hours spent in front of the mirror (and associated dollars spent on beauty products) add up. This is time that could be spent doing other things, like getting ahead at work, taking up a new hobby, or maybe even sleeping.

So even though these standards, and wider societal attitudes, have long-standing roots, this doesn't mean we shouldn't question them. It's vital that a young woman's sense of self-worth is based in something other than the way she looks. And women who don't abide by these expectations shouldn't be penalized.

But a new double standard has developed more recently, wherein beauty rituals that uphold femininity (like shaving one's legs and wearing a push-up bra) are criticized and frowned upon, but any steps to appear gender-nonconforming (like wearing men's clothing and dying one's hair unusual colors), which may be just as time-consuming, are celebrated.

Telling women there is only one way to behave, whether it's abiding by expected beauty routines or rejecting them, is equally unhealthy. If women are truly free, they should have the freedom to choose whether they want to spend extra time getting ready in the morning or not, and whether they want to appear feminine or in a more gender-nonconforming way. As someone who prefers masculine clothing, I rarely wear dresses, but I don't think putting one on makes a woman any less liberated.

There is no reason why a woman can't be intelligent and critically minded while also caring about the way she looks. If anything, it's misogynistic to assume that a woman can't be both. Interestingly, it's often women who police each other's appearances.

A World Without Men

Men and "the patriarchy" are frequently blamed for women's desire to be beautiful. Another critical influence, however, rarely addressed in this debate, is the pressure exerted from other women.

Women are in competition with each other for sexual partners, a phenomenon known as *intrasexual competition*. Tactics that increase the likelihood of successfully attaining a desired mate include everything from self-promotion to indirect aggression toward other women, including putting down female rivals, insulting their looks, and calling them promiscuous.[7] Furthermore, tactics like encouraging women to go makeup-free have allowed females to implement these competitive strategies under the guise of advancing women's rights. It is the same mentality underscoring some women's disdain for plastic surgery. At a time when women are being encouraged to lift one another up, champion the sisterhood, and bond over the rejection of beauty standards, what it really is is a subconscious way to cut out one's competition.

Contemporary feminism places pressure on women to eschew traditional femininity and basically let themselves go physically in order to reach enlightenment. For those who are unwilling to do so, they are seen as victims of societal brainwashing. A more pertinent question is who is truly benefiting from these changes.

We only have to look to the multibillion-dollar modeling industry to see the perks of being an attractive woman. It is a profession that is unconventional in that the fairer sex far outearns men. You would think an industry in which women make 75 percent[8] more than their male counterparts would garner some praise instead of being considered demeaning to women.

At the same time, women who are critical of the modeling world will applaud men who wear makeup and high heels. Why? Partly because liberal feminism has hailed atypical gender expression as more socially acceptable than that which is typical, but also because, regardless of whether these men are gay or straight, these women aren't competing with them for sexual partners.

It is a positive stride forward that women are being encouraged to support one another instead of falling into the trap of dragging each other down. But it's a straight-up lie to claim that a world without men would be a female utopia rid of cattiness, gossiping, and female competitiveness.

I have seen op-eds going so far as to advocate against heterosexuality so that women will be completely sufficient without men. Putting aside the fact that this would eventually lead to the extinction of our species, women would not be better off for it.

Masculinity has been unfairly pathologized. (We touched on "toxic masculinity" in Chapter 3.) Sexual and physical violence are viewed as extreme forms of masculinity, when this is completely unfounded.

Successful mating consists of attaining a partner with good genes so that one's offspring will survive and be successful at passing on their lineage. Men's behavior is, to some extent, the result of female sexual preferences. If women didn't want to mate with masculine men, these traits would have been removed from the gene pool long ago.

It's a case of the lady doth protest too much. "Toxic masculinity" is the result of women's sexual preferences over thousands of generations. Contemporary feminists are punishing an entire generation of men for the mating preferences of their female ancestors.

A study in the *Personality and Social Psychology Bulletin* found that women prefer men who exhibit benevolent sexism over those who do not, despite also finding these behaviors undermining and patronizing.[9] Benevolent sexism consists of chivalrous behavior, such as opening doors and paying for dates, and positive attitudes about women that reinforce the belief that women are less competent, relegating them to traditional gender roles. (This is in contrast to hostile sexism, which involves overtly misogynistic views.) Regardless of whether women endorsed high or low levels of feminist beliefs, they interpreted signs of benevolent sexism as signals that a man is willing to invest in them through an ability to provide, protect, and commit to them. The positives associated with this behavior outweighed the negatives.

From a brief survey of my straight male friends, the more progressive and feminist a woman proclaims she is, the more she prefers dominant, masculine men. It brings to mind one of my former, very feminist female friends who, despite protesting how destructive masculinity was, somehow always ended up dating men who were built like linebackers, who looked like they could probably snap a person in half during sex.

Casual Sex and Courtship

One of the greatest myths perpetuated about female sexuality is that women enjoy casual sex as much as men. Young women are encouraged to pursue hooking up because that is what empowered women are supposed to. Those who abstain are seen as unfortunate casualties of a culture rampant with slut shaming and repressive gender norms.

Considering everything we know about evolutionary psychology, this doesn't make much sense. Despite feminism's incessant prodding that women indulge in no-strings-attached sex, men consistently report greater enjoyment of it, while women experience more negative reactions. After having casual sex, women are more likely to self-report depression, regret, and feeling "used." Men, conversely, self-report feeling like they had used their female partners.[10]

Is there any truth to the idea that social standards influence this behavior? When we look at sociosexuality (which refers to a person's enjoyment of sex without commitment), in more egalitarian societies, like Norway and Denmark, women enjoyed casual sex more, but so did men, leading to larger sex differences.

This is not to say that women should cease having casual sex. Women should be free to have as many libidinous encounters with as many different partners as they'd like, if that's what they choose to do. I'm a strong believer that what's good for the goose is good for the gander; if promiscuity is socially acceptable for men, then it should be socially acceptable for women, too, and conversely, if it is frowned upon for one sex, it should be frowned upon for both sexes. There definitely still exists a double standard regarding women's sexuality—for example, that men are free to roam sexually while women must remain chaste to preserve their inherent value as future wives and human beings, more broadly—and even though these attitudes are rooted in evolutionary psychology, that isn't an excuse to hold women to a different standard than men.

A person cannot make good decisions without having all of the information available to them. Currently, only one message is being aimed at women in the mainstream, and unless a young woman has parents, educators, or health professionals in her life—who have not embraced this particular ideology around casual sex—whom she feels she can turn to to talk about these very intimate and in some cases, potentially embarrassing, topics, the only place she will be able to access information will be on the Internet and through exposure to messages in the media.

There are, of course, many women who do enjoy casual sex, and many men who don't, and so long as everything is consensual, it's no one's place to judge. My concern is that only one perspective is being promoted as acceptable for young women, once again dictating to women that they should behave like men in order to be well adjusted, yet paradoxically, young men are being made to feel shameful if they behave like men. If a woman is more interested in having committed relationships, that should be completely fine, too. She should not be made to feel as though she is uptight, unenlightened, or a sexual prude.

The narrative that sexism and socialization are the most powerful influences on women's sexual behavior—or lack thereof—leads women astray in several ways. Young women reach out to me, asking for sex advice, because in many instances, what they find in the mainstream conflicts with their own experiences with men.

First of all, male sexuality is concordant. This means when a man is turned on psychologically, he's also turned on physically, down below. Female sexuality, on the other hand, is much more context-dependent, and is lower in sexual concordance. Women may be psychologically turned on, but not physically, and vice versa. It can take a bit of extra time for a woman to get warmed up. From an evolutionary perspective, this ties into women's greater selectivity when choosing their mates.

For instance, a woman's brain processes sexual cues differently, depending on where she is in her ovulatory cycle. A recent study using functional MRI demonstrated that women possess increased efficiency in inhibitory brain function when they are potentially fertile.[11] This is adaptive because it allows a woman to be more cautious when evaluating a potential partner at times when it is more likely that she will become pregnant.

It's not appropriate or helpful to use male sexuality as the standard, because women's bodies don't operate this way. Women who believe female and male sexuality are no different will feel pressured to perform similarly to their male partners in the bedroom. When they don't or can't, they will internalize this, thinking there is something wrong with them.

It may be the twenty-first century, but a man should still be required to make the first move. Initiating contact or romantic interest is a small act demonstrating he is willing to invest in her. As a woman, you do not want your boyfriend to treat you like one of the guys, even if that sounds commendable in the name of gender equality, because doing so requires him to put in much less effort. You want to know that if and when the day comes that you're heavily pregnant and immobile on the couch, he will bring you what you need to ease your discomfort, instead of muttering that you should get it yourself. Even if you have no plans to have children, you want to ensure that your partner will prioritize you.

I can feel some of you giving me a heavy side-eye, but hear me out. To remove the fourth wall for a moment, *I* can't even believe I'm giving this kind of advice. My younger self would be shocked and mortified, much like my two male friends at the start of this chapter were. If it's empowering for women to go after what they want in the corporate world and every other aspect of their lives, why shouldn't the same standards apply in dating?

The answer: because we have not yet severed our modern-day sexual behavior from its history. It's a system that has served us quite well, considering that we exist today. Acknowledging as much doesn't mean women will get the short end of the stick, nor does it require perpetuating the cliché that women should be sexually submissive and men should be sexual aggressors. If anything, this knowledge can be used to the advantage of both sexes.

Women have been told to take charge in their romantic lives as a sign of independence, confidence, and female strength. While I don't think women should be sitting around, waiting for a guy to send a text, there is something to be said for letting him come to you.

While at a barbecue a few years ago, I witnessed an exchange between two young people that speaks to this. Seeing someone you find attractive is always a bit of a balancing act. You want to give signals that you are interested, but not *too* interested—sexually appealing, but not too forward. Finding the right words to start a conversation can be hard, especially if it's with a stranger.

Her choice for an opening, I kid you not, was "Men are trash, don't you think?" as though this was a nonchalant way of flirting. The guy in question looked so perplexed, I actually felt sorry for him. They made a bit of uncomfortable small talk before she promptly put her number in his phone and turned and walked away.

My first piece of advice would be, don't do this. As socially acceptable as male bashing is nowadays, what kind of man is going to agree with you? If he is even moderately well adjusted, alarm bells will be going off in his head, sounding a warning that a woman who thinks men are subhuman is not a woman he wants to date.

Second, it takes courage to approach a man and give him your phone number, so I give her full credit for that. What's the downside? You won't actually know if he's into you. Since women enter romantic situations at a different baseline investment than men, there needs to be some recalibration to offset that disparity. Requiring men to, at minimum, initiate interest will help to weed out those who are just going to waste your time. It also means that he values you.

A man who isn't required to invest will almost by default be interested. Why wouldn't he? It's potentially free sex!

Is it old-fashioned to think this way? Probably. But if it gets you what you want and helps you avoid being heartbroken, what's wrong with that?

Beyond that, I'd advise women to not play hard to get. If you are interested in someone and they ask you out, say yes. Don't expect men to be persistent in the face of repeated rejection, because the line between adorably persistent versus creepy and stalker-like is very fine, and in our current climate, a man is not going to risk falling into the latter category, no matter how much he likes you.

As for men, don't apologize for being masculine. Pull your weight in a relationship and be gentlemen. Some women may find holding doors open and gestures of the like unnecessary, but they shouldn't be *offended*. I've had male friends tell me progressive women sometimes get angry at their acts of chivalry. In return, my friends have stopped putting in those efforts when on dates with other women. I'd say, if a woman gets mad at you for something that was well intentioned, take it as a sign that this is probably not someone you want to be with.

Truly successful relationships require a healthy and accurate understanding of not only one another, but human behavior; otherwise you will end up with a partner who is only telling you what you want to hear. I can draw from an example in my own life.

Studying male sexuality has come with an incredible bonus that I hadn't anticipated at the time—being able to accurately tell whether a man is being honest about his sexual behavior and history, even when I'm not asking these questions as part of a formal research protocol. I was once on a date with a man who told me, with a straight face, that he never looked at pornography when he was in a relationship because he stopped finding other women attractive.

If I still bought into the lie that men and women are a facsimile of one another, and men become as emotionally invested as women do after sex, I might have thought I had lucked out and found the most romantic person on the planet, someone who quite literally only had eyes for his one true love. But because I understand human sexuality and men's on-average greater preference for novelty and visual input, I found this random proclamation about his porn-viewing habits unnecessary, off-putting, and insincere.

Some people may actually *want* a relationship in which their partner consistently tells them sweet nothings. This may be exactly what they're looking for, especially if they are doctrinaire about their political—or sexual—beliefs and one of their dating requirements is that their partner agrees with them in every way. I encounter many men who are more than happy to signal their status as "feminists" and "male allies" who want to "fight systemic oppression" and "overthrow the patriarchy."

I've spoken at length about my disdain for male feminists due to this type of posturing. Almost consistently, they turn out to be master manipulators and abusers of women, as we've seen with many high-profile cases in the media, such as convicted sexual abuser Harvey Weinstein. Feminism tells women that good and decent men should behave like women, so feminist women in turn prefer dating men "like them," when these men are acting duplicitously and repeating by rote what they think women want to hear.

Because self-proclaimed male feminists know I've been critical of feminism, they have no problem telling me what they *really* think. Some consider feminist beliefs to be irrational, and expect women to be less competent than men. One individual told me he subscribed to evolutionary theory, but would never dare tell his girlfriend this—instead, in her presence, he agreed it was a figment of white men's imaginations. These men only say such things in order to score points with women and secure sexual partners.

It's one thing to be in support of equality for women, as I believe all men should be. It's another to show up at women-centric events wearing a pink vulva-themed hat and a giant badge that says, "I'm an Ally," while extolling the virtues of "unpaid emotional labor" and inserting uplifting yet irrelevant quotes from feminist thought leaders into everyday conversation.

So, ladies, if you are in the market to date a sane person, you should hope a man doesn't agree with extreme-left feminism. Because in the long run, it will get old very quickly. People can't hide who they are or what they really think forever.

The Counterargument

While writing this chapter, one of my concerns was that some of my arguments could be used to justify the distasteful theories of misogynistic men's rights movements. Although it's important to note that not all men advocating for men's rights are misogynistic, a peek into associated meetings and online forums reveals that many view women as "entitled bitches" (their words, not mine) who are only seeking to use men at their disposal. According to movements like MGTOW (Men Going Their Own Way), the world is "gynocentric" and women have unearned "privilege" due to being propped up on a pedestal by that "corrupt" feminist movement.

There is some overlap between the men's rights and the "incel" (involuntary celibacy) communities. Incels are an online subculture of men who are frustrated at their inability to obtain sex and genuine affection from a romantic partner. Some within the community go as far as advocating rape to gain sexual gratification from women.

Certainly, someone who hates women, regardless of whether he identifies with either of these social movements, can use evolutionary psychology's points as a justification for his opinions. I understand why some women, in return, detest evolutionary explanations for behavior and any evidence for sex differences, as a general rule. But denial of these truths only weakens the argument for gender equality. If a person doesn't fully understand the evidence, they cannot argue why it isn't acceptable to extrapolate sexist ideas from it.

On the flip side, this chapter will also likely enrage feminists who believe evolutionary psychology portrays women as weak, submissive, and dependent on men, and that men who abide by gendered norms, demonstrating chivalry and wanting to protect and provide for their partner, are condescending. Indeed, there are some men who are benevolently sexist, who see women as less capable, and that is why they make these sorts of gestures toward them. But not every masculine, chivalrous man thinks this way.

Nothing about evolutionary psychology offers support to either of these fringe ideologies, and in fact, both perspectives are two sides of the same, spiteful coin. It's truly a shame that, as a scientific discipline, evolutionary psychology isn't mandatory reading as part of university curriculum. It would undoubtedly help men and women navigate not just dating and sex, but understanding their own attitudes and behavior, to have more fulfilling sex, and win the ultimate prize of falling in love.

MYTH #8

GENDER-NEUTRAL PARENTING WORKS

One of my colleagues once told me a funny story that illustrates biology's tenacity in the development of gender. While traveling abroad and staying with family friends, he learned the couple was raising their twins, a girl and a boy, with the goal of being gender-atypical. The girl had been given boy-typical toys, like fire trucks, to play with. Similarly, the boy had been given girl-typical toys, like dolls.

My colleague suffered from insomnia at the time, and when everyone had gone to bed, he would find himself lying awake in the living room. Every night, he would observe the most remarkable thing—the children would sneak into each other's room to play with sex-typical toys. The girl preferred her brother's dolls and the boy preferred his sister's trucks.

We've all seen at least one news story by now about parents who don't make any assumptions about their child's gender. These parents are instead waiting until the child is old enough to tell them which gender they are. Some will go as far as allowing a child to pick their pronouns, sometimes switching back and forth between she/her, he/him, and they/them. In the minds of these parents, the sex and gender "assigned" by the doctor has no relevance to their child's internal sense of who they are.

As legal recognition of nonbinary gender status on government-issued documents grows in availability across countries like the United States, Canada, and Australia, this way of thinking has become ubiquitous in childrearing. The gender-neutral parenting approach (which includes giving children gender-neutral names, toys, and clothing) has expanded in recent years to spawn an increase in the number of parents raising gender-neutral "theybies." In response, nonbelievers roll their eyes and shake their heads, calling it "child abuse" and "radical leftist parenting." Many blame parental psychopathology and mark these changes as a sign that society is doomed.

The rationale for this trend is the mistaken belief that gender-typical traits and interests are taught to children from the moment they are born, and unless children are raised in a culture void of gender, they will be boxed into fixed categories, reaching only a fraction of their potential. Boys are assumed to be natural-born leaders, falling prey to their own aggressive inclinations, bottling up their emotions until they explode. Girls face limitations through assumptions about being too emotional, irrational, and not fit for responsibility.

It seems everyone has fallen down the gender-neutral rabbit hole. Children's Hospital Colorado removed any designation of biological sex from its wristbands in order to be sensitive to the needs of children who identify as the opposite sex or a third gender, stating that this would not affect patient safety.[1] And Planned Parenthood announced that reproductive anatomy is not "male or female," with the goal of fighting discrimination against transgender people.[2]

During Barack Obama's presidency, the White House hosted a toy-sorting event in which the president made a point to put sports toys in the "Girls" bin. Sweden opened two gender-neutral preschools, wherein children are referred to using the gender-neutral pronoun "hen," instead of boys and girls. North American parents have similarly taken up the cause, refusing to refer to a "boy" or a "girl," referring instead to a "child" and another "child." My home country's national treasure, Céline Dion, released a gender-neutral clothing line for children. The Brit Awards considered removing male and female artist categories to make room for musicians who identify as nonbinary.

Everything from gingerbread cookies to sewer coverings to Santa Claus has undergone a gender-neutral rebranding, but the debate is full of ideological inconsistencies. For example, in gender-typical children, gender is denigrated as an irrelevant construction that can and should be unlearned, but when it comes to gender dysphoric kids, it should be prioritized over biology. Similarly, those who say gender is a social construct and femininity and masculinity are learned will use gender stereotypes to justify why a little boy who wears dresses, or a little girl in overalls, is really the opposite sex.

You can raise a child without saying "this is for boys," and "this is for girls," or conversely, treating them as though they are amorphous, genderless beings. Instead of calling it gender-neutral parenting, why not just call it being open-minded?

Is Gender Harmful?

Every holiday season, parents, grandparents, and anyone with children in their lives are bombarded with a constant stream of cultural messages reminding them that gendered toys should be avoided like the plague. Aisles that once segregated toys by sex and soft washes of pink and blue have been sterilized and recategorized by age and interest. In 2015, after a number of parents complained on social media, Target stopped sex-segregating its toys. Later that year, Disney followed suit, as did Mattel in 2018. Nowadays parents will rarely encounter toys labeled as being for "girls" or "boys."

Children's bedrooms and play areas have become another casualty. Color palettes consist of strictly neutral tones and greige or carefully vetted primary colors. Wallpaper themes in a child's bedroom forbid fire trucks, cute animals, and fairy-tale allegories because they enforce the gender binary. I am all for letting kids be the masters of their destiny, but I can't help but feel like these children are being deprived. Previous research has in fact shown that female-typical toys, like dolls, are associated with more complex play and that both girls and boys can benefit from having access to them.[3]

The idea that masculinity and femininity are learned is one of those myths that just won't go away. Gender is not forced upon children by way of parental messaging, teachers, culture, or the media. Even if children absorb gender stereotypes, this won't necessarily translate to self-limiting baggage they will carry with them throughout their lives.

Once upon a time, parents preferred a gender-conforming child. Girls were supposed to be dainty princesses and boys were to be stoic adventurers. Allowing a child to play with gender-atypical toys elicited fears that doing so would turn them gay. (Let me reiterate, there is absolutely nothing wrong with being gay.)

Now, a polar-opposite trend has emerged, with a preference for gender-atypical children. Thankfully, it seems only a minority of parents have opted for such a preference, but it is steadily growing. A curious Web search will return a plethora of think pieces and media reportage of parents priding themselves on their tolerance. In almost every story, the family in question has somehow managed to raise multiple children who were gender-nonconforming. Parents enroll their boys in ballet class and girls in rough sports like football in an effort to encourage the widest range of interests possible. But pushing otherwise gender-typical kids to pursue gender-atypical activities is the same line of erroneous thinking evident in parents who think giving gender-atypical (or gender-typical) toys to a child will alter their sexual orientation. A child will naturally gravitate toward whatever they're interested in due to predetermined biological influences, regardless of whether they are gender-conforming or nonconforming.

I'm sure many parents will be repulsed to read such a thing. Therein lies the double standard: If a child is gender-nonconforming, this is interpreted as biological and something that shouldn't be dissuaded or tampered with. But if a child is gender-conforming, this is seen as the result of social influence and something that parents should actively try to change. I often see boys who are gender-atypical, allowed by their parents to express themselves in a hyperfeminine and in some cases, inappropriately sexualized way, pouting with duck lips in photos and posing seductively. In the case of child drag queens, for example, little boys—some as young as age eight—perfect their makeup and hair and put on skimpy outfits to gyrate to, in many cases, explicitly sexual songs onstage. As someone who spent more nights than I can count in drag clubs with my friends when I was younger, I fully support young kids, especially feminine boys, expressing themselves. But I find the hypocrisy mind-numbing—would the adults cheering on drag kids allow their daughters to pose in the same way?

I understand why gender-neutral parenting is attractive to parents. There is a desire to prevent daughters from being passive and self-sexualizing, and to equip sons with good language skills and the ability to express their emotions. Parents *should* want to give their child the best life possible.

They should also have faith that even if a gender-atypical child is put in an environment where they are forced to conform to gender norms—whether it's due to teachers or classmates or society—they will return to doing as they please as soon as they are able to. A little boy will begrudgingly play with tool kits if he must, but he will always prefer dolls. When he gets older and has the freedom and autonomy of adulthood, he will gravitate toward occupations and interests he truly enjoys, which will likely be female-typical, like hairdressing or social work. The same can be said for atypical little girls and the desire for male-typical roles.

Some parents voice concerns that their child may be intersex and the doctor may have unknowingly "assigned" the wrong sex at birth (see Chapter 1). These parents fail to realize how rare intersex conditions are. There is a 99 percent chance that, upon reaching adulthood, their child's sex will be the same as what everyone thought it was. Intersex people deserve the same rights and protections as everyone else, but it isn't unreasonable to think that the majority of kids will identify with their birth sex.

In many cases, I'd reckon it has more to do with a parent wanting to signal how accepting they are. The same can be said for those at the start of this chapter, who are unwilling to reveal the sex of their baby.

Gender-neutral parenting was once about the toys a child wanted to play with and what they wanted to be when they grew up. It has since expanded to encompass social justice missions like "smashing the gender binary," including questions about identity and who a child believes they are. We see this with the trend of children as young as age three, who, according to their parents, identify as nonbinary and use pronouns like "they/them." There will always be parents who maintain a desire to stand out as different and special.

I am 100 percent in favor of supporting gender-nonconforming kids, but there's no reason they need to be categorized as another gender. Developmental research has shown it isn't until roughly age seven that a child comprehends that surface changes to one's appearance or behavior do not alter their gender. Until then, many children won't understand that a little boy who puts on a dress and a wig is still a boy, even though he looks like a girl. If anything, parents who refrain from correcting a child about their gender will only exacerbate this confusion. Does a child have the cognitive capacity to identify as a third gender? Much like what I've said about so-called transgender kids, the answer is no.

The Influence of Biology

I wish I could tell every parent out there that biology isn't something that needs to be feared. Parents have been told that if they don't offer a perfectly balanced environment, they are restricting a child's potential—little girls will be deprived of occupational opportunities and little boys will be emotionally stunted.

The science in favor of biological explanations for gender is uncontested. As I mentioned in Chapter 2, social influences can affect the extent to which a person's interests and behaviors are expressed, but they cannot override the underlying preferences themselves. Gender is dictated by prenatal hormone exposure, as opposed to coercive gender norms imposed upon infants the minute they exit the womb.

Social markers for gender may change as decades go by, but this doesn't mean children are socialized into having a gender. For instance, men wore ruffles in the 1700s and boys in the Victorian era wore dresses. Blue was once considered to be a feminine color and pink was considered to be masculine.[4] This doesn't disprove that gender is biological, only that the expression of gender changes depending on what is considered male- and female-typical.

A typical boy is exposed to high levels of testosterone, and when he is born, he will gravitate to mechanically interesting activities, like playing with wheeled toys, and related occupations in adulthood. The same can be said for girls who experience high levels of testosterone exposure. As shown in girls with congenital adrenal hyperplasia, even if their parents give them more encouragement for playing with dolls, they will still prefer toys typical to boys.[5]

This is because greater exposure to testosterone in utero is associated with male-typical interests, and as we saw in Chapter 2, something called the "extreme male brain." The extreme male brain tends to correlate with higher efficiency in systematizing and in some cases, a diagnosis on the autism spectrum. Brains that are exposed to lower levels of testosterone, conversely, are more efficient at empathizing. (This is not to say, however, that people on the autism spectrum have deficits in empathy. People with autism may have difficulty being attuned to others' emotions, but they are capable of feeling and responding to them.[6])

Girls show a preference for socially engaging activities and occupations. This difference between children regarding preferences for people versus things is detectable within the first two days of life.[7] Baby girls preferred looking at their caregivers' faces and baby boys preferred looking at mechanical mobiles. Babies as young as nine months old have shown gender differences in the toys they choose. Again, girls preferred playing with dolls and boys gravitated toward trucks and cars. This was before they're able to even recognize gender as a concept, something that generally happens at around eighteen months to two years old.[8] The denial of biology is so strong, however, that I won't be surprised when critics soon say that gender stereotypes are learned before birth.

We also see the same behavior in our primate relatives, including vervet[9] and rhesus monkeys.[10] Despite lacking socialization from their caregivers or other monkeys, young female monkeys will choose dolls, and male monkeys will choose wheeled toys, similar to what is observed in human babies.

Critics of these findings frequently criticize the use of animal models, saying it isn't appropriate to extrapolate the results to human beings. But researchers are unable to do experiments using human participants because it would be unethical. Can you imagine the outrage if scientists wanted to manipulate the prenatal environment in a mother's body and see what happens when the child is born? Do you think researchers could get away with changing the hormonal profile of one randomly selected group of fetuses and monitoring another set as the controls? Ethics committees, designated for the purposes of putting the kibosh on poorly designed and potentially harmful research, would not allow it for a second. A perfectly designed study, using humans in order to draw the most certain scientific conclusions, could only exist in science fiction.

A very well-known case study illustrating this pertains to a Canadian man named David Reimer. A physician named John Money surgically reassigned Reimer as female after he lost his penis at a young age in a botched circumcision. Money believed that Reimer could be successfully socialized and raised to live his life as a girl.

Upon reaching adolescence, however, Reimer felt something was amiss, and returned to living as a male. At the age of thirty-eight, he tragically took his own life due to these early childhood experiences and a number of difficult life circumstances he was weathering at the time.[11] Reimer's case demonstrated that gender is innate; whether we feel female or male is not learned.

A study published in the *New England Journal of Medicine* followed sixteen boys who were reassigned as girls due to a condition called cloacal exstrophy, which caused some of their lower abdominal organs to be exposed at birth, as well as malformations in the penis.[12] Upon growing up, eight of the children had returned to identifying as male and only five continued identifying as girls. (Regarding the remaining children, two were unclear about the sex they identified as and one preferred not to disclose this information.)

Where, you might ask, is this anti-biology trend coming from? My guess is that many of the most ardent supporters of the genderless trend, or the denial of sex differences, are women who have experienced discrimination based on sex in their own lives, who were held back and unjustly lost opportunities they wanted.

This is not to say that sexism and discrimination don't exist or that children aren't streamlined as a result of their sex. Gender-atypical girls do face bullying and teasing for taking part in boy-typical activities, and the same can be said for gender-atypical boys who are intent on doing the things that girls do. But children deserve more credit than they are being given. They are tenacious and resilient, and a bit of resistance will not deter them.

What Parents Can Do

So many parents have asked me what sort of an approach they should take when raising their kids. For parents who are concerned about inadvertently limiting a child's potential, the bottom line is offering choices without being dogmatic in either direction. I'd suggest exposing kids to a variety of activities and ways of expressing themselves that are gender-typical and atypical, without making any assumptions about their inherent interest or ability. Let boys—and girls—wear pink ball gowns or short haircuts and play with whatever toys they want.

In the event that your child adheres to stereotypical gender norms, don't panic. A greater emphasis should be placed on being loving and supportive instead of scrutinizing gender roles. Redirecting a child's play toward a more gender-exploratory, nonconforming course is just as invasive as forbidding them from deviating from traditional gender norms.

What will happen to these genderless children? I hear from countless parents who recall their experiences raising daughters in a gender-neutral way. They purchased only boys' toys, like sports gear and miniature cars, for them to play with. But then the inevitability of the "princess phase" hit. Their daughters encountered dolls at a friend's house and now wanted to play with their own.

What's wrong with that? The parents are perturbed. In one parenting blog post, a mother wrote about how, despite family members pushing dolls and dresses on her four-year-old daughter, the girl chose a black bicycle instead of one that was pink as a birthday gift. The woman hoped this might be a sign of victory. In another post, a sociologist described attempting to ban princess-themed activities and toys from her own childrearing practices, but eventually relented, believing that social factors, like a child's peers and sex-stereotypical advertising, would be more influential than their parents. I have indeed heard of parents prohibiting female toys in the home, replacing a girl's collection of stuffed animals and ballerina tutus with wooden building blocks and black-and-white photos. Yet these parents would certainly lavish a son with the exact same arsenal of girly toys.

One reader recounted a story of giving their daughter's classmate a doll at her birthday party. The classmate's mother explicitly forbid it, sending the reader's daughter back home with the doll in tow.

I once read of a family who proudly and publicly announced the naming of their daughter after a prominent computer programmer. They were then devastated when the girl came home from school one day, announcing a preference for playing with dolls. I couldn't help but laugh a little at the thought that biology was trolling them.

In other cases, mothers will reach out to me to share their experiences. Some women, who identify as feminist, tell me they went to painstaking lengths to ensure their child was raised in a gender-neutral home, that no gender norms were enforced, and that girls, in particular, were treated no differently from boys. It didn't take very long after the child was born for these women to realize that they really didn't have much say in the matter. Children had their own minds about it, and girls and boys were very different.

No matter how much freedom a child is given, most will, time and time again, pick out gender-typical toys to play with and exhibit personality traits typical of their sex. These mothers would tell me that feminist ideas about gender were great in theory, but they didn't materialize in real life. Much like my own awakening that began with the questioning of social constructionist ideas about gender, they too started asking themselves which other tenets of feminism were flawed.

Parents who are not ideological will realize the power of nature in these situations. Those who are ideological will think they failed at gender-neutral parenting and that they should just try harder.

Gender-neutral parenting is not a reason to stop children from pursuing what they want. If you try to force kids to play with opposite-sex toys when they prefer toys typical to their sex, they will be bored, or alternatively, will get creative. I've had parents tell me their boys, upon being given dolls, will swing them around mercilessly by the hair as though they are a weapon. Girls will arrange toy trucks into a family and tuck them into bed.

For those who are concerned that girls are being force-fed a life of femininity, I would argue that a girl who is gender-atypical will turn away from these cues if she's truly not interested. For gender-nonconforming girls who are bombarded with societal messaging about femininity through movies and online games, they will say, "I don't want to watch this," or "I don't want to play this," and "Where are the ones for boys?" Gender-nonconforming girls don't need the world to tell them to be nonconforming; they will already be that way, from the moment they are born.

How would I know? I have always been male-typical, and to this day, despite looking very feminine, I still feel much more masculine. As a child, anytime I wore a dress, I remember feeling as though something was very wrong. My childhood friends were all boys because I was more interested in play fighting (and winning) than putting on lipstick and organizing tea parties. I used to think this was because my family was open-minded when I was growing up, but upon learning about biology, I now believe differently.

Some ask why I seem to be arguing against a movement that

supports kids like the one I was. I'm about scientific data and facts. Just because I was and am different doesn't mean everyone has to be.

Nevertheless, some adults seem to think they can program a child's destiny. Gender-neutral names, like "Skyler" and "Avery," are growing in popularity, but what is the underlying message associated with them? It is almost always a case of parents giving their daughters masculine-sounding names, as opposed to the other way around. This would appear to be in line with the same, insidious form of sexism—pushing girls to be more male-typical or gender-neutral, or at the very least, less feminine. This does nothing to promote the treatment of girls, regardless of gender expression, as equals.

Gendered stereotypes are not in themselves harmful, and it is ill-fated to impose this trend on children and society, more broadly. Girls face a constant pressure to behave more like boys, and vice versa, and it is female-typical occupations that are undervalued. There is no reason why a feminine girl can't grow up to be a CEO, and if she'd rather do something else with her life besides climbing the corporate ladder, that should not be frowned upon. Why should a woman's value be defined by her success in a male-dominated field?

It is unacceptable for women to be denied educational opportunities or to be held back in the workplace because they are female. But many women do not want to sacrifice their personal lives in the name of a high-income, high-status career.

I will never forget the poignant words of advice an assistant professor once gave in graduate school. Her efficiency as a scientist was unparalleled, as she managed to pump out research publications nonstop. For most female academics who choose to start a family, there will be a gap in their curriculum vitae due to time taken away from work to give birth and raise the baby. With this professor, however, you would not have known she had two children. Immediately after giving birth to her second child, she was right back in the lab, working on another study.

She once gave a talk to eager grad students about how to manage a successful work-life balance. "If you want to succeed in your career, you have to sacrifice your personal life," she began. "If you want to succeed in your personal life, you have to sacrifice your career." Then, with a look of pallor, she added, "If you want to succeed at both, you have to sacrifice *yourself*."

It was dire but brutally honest. Women, in most cases, face greater responsibilities with regard to childrearing in addition to balancing a busy career. Several students came out of that lecture with the realization that they weren't willing to sacrifice themselves if they wanted children, so something else would have to give.

If we want the world to be more accommodating to women (and we should), we need to stop pretending that the sexes are identical. Telling ourselves that sex differences are inherently meaningless hinders progress that would actually help women.

We see few women in male-dominated fields like construction, manufacturing, and coal-mining, but no one is pushing for gender parity in them. In areas where women are dominating, such as college enrollment, there fails to be a similar level of concern around increasing the number of men.

The solution is not to replace one set of rigid and extreme views about gender roles for another, whether it's traditional ideas or far-left progressive ones. In Chapter 9, I discuss what parents can do to counter this misinformation.

MYTH #9

SEXOLOGY AND SOCIAL JUSTICE MAKE GOOD BEDFELLOWS

I knew sexology was in trouble the day I encountered the male gaze.

The confrontation wasn't in the way that you probably think. The "male gaze" was coined in 1975 and remains a key concept in feminist film theory.[1] It refers to the sexual objectification of women, particularly in the realm of art; that is, the camera gaze is implicitly male, operating from the perspective of straight men ogling women.

I recognized the term from my days as an unwavering feminist. I had never heard it used in a research setting, however, and certainly not during a scientific talk. At the time, I had just finished my PhD, and a friend who knew of my research interests had forwarded me a flyer for the event. It sounded like it would be right up my alley, and after wrapping up my interviews that day for an upcoming column, I popped across town to attend the presentation at a nearby campus.

The researcher was presenting his findings regarding sex differences in hormonal response. On his first slide, he showed that, at baseline, women's cortisol levels were higher than men's.

Someone in the audience had a question. "Is this because of the male gaze?"

It took me a second to fully register what had been said. *Did someone really just ask about the male gaze, a completely ideological concept, during a scientific presentation?*

Indeed, someone had, as the researcher responded something along the lines of "It could be." And he didn't seem to be joking.

Sitting in the front row, I turned 180 degrees to look at who had just posed the question. The room was small and silent, and as I turned around, it became clear to everyone what I was doing. I knew I couldn't be the only one completely aghast at the question. But I was the only one who was willing to have my thoughts known.

Is this really happening?

I thanked my lucky stars that I had made the decision to self-deport from academia. If I were still in research, I too would have felt the pressure to avoid making a fuss, and probably would have stayed, facing forward, eyes obediently glued to the front of the room.

This occurred in 2017. It was only the beginning of what was to come.

Activist Science Is Not Science

There is activism and there is science. Activist science, no matter how passionate or well intentioned, is not science.

Activism has no place in scientific research. There is no such thing as "feminist science," "queer science," liberal or conservative science, or what have you. If you're doing the scientific method properly, it really doesn't matter what your politics are. And yet themes of "eliminating injustice" and "supporting the resistance" crop up in the least expected of places.

A few years ago, in Toronto, I attended a protest that was organized in the name of defending science. As I walked around the protest grounds, I could not believe the number of people who were vehemently rallying against President Donald Trump's climate science denial, but had nothing to say about left-leaning denial of biology. One of the speakers gave an entire speech dedicated to the underrepresentation of women in STEM, blaming societal factors for the imbalance in the sex ratio.

When I approached her afterward to ask what she thought of the enormous scientific literature supporting biological explanations for sex differences in occupational interest, she told me research didn't yet know "what is nature and what is nurture."

Really, I shouldn't have been surprised. The tentacles of intersectionality were loudly and proudly woven into the event's core principles. Much of the content that day blamed capitalism, fascism, and "power structures" for whatever it is they considered antiscience.

In theory, social justice and its efforts to end discrimination are a good thing. But like many political movements, it has been hijacked by extremists who care very little about what the movement initially stood for.

Illiberalism

People often ask me what the fallout has been since I became a journalist. I have lost friends over politics, friends who couldn't understand why I didn't hold the views a "good liberal" should. If you're left-leaning, but don't agree with every aspect of progressive thought, there's a good chance you will similarly find yourself disowned.

I, however, remained blissfully unaware of this throughout the majority of my time working in research. It was only after I published that first op-ed that I had my awakening. So many of the things I once believed were not true. The problem ran much deeper than the glorification of transgender kids. I soon realized that others had had a similar awakening, only much sooner.

It was a Friday evening after a long week, and I sat on my sofa, debating whether to go out and see my friends like a normal person, or continue merging with the upholstery. My phone suddenly lit up.

"Hey Debra," it read.

It was my friend Tom, from graduate school. His lab had been across the hall from mine and we often took breaks, visiting each other, when we had been staring at our respective computer screens for too long. I was on my way to graduating and he had finished his degree the year before.

I picked up my phone and we typed back and forth, catching up, when his messages suddenly took a serious tone. He told me he'd always been different from everyone else. There was something he had to tell me.

An instantaneous realization flashed through my mind.

Tom is gay, I thought.

Growing up in the gay community bestowed upon me impeccable gaydar and the uncanny ability to tell someone's sexual orientation within five seconds of meeting them. Tom didn't strike me as gay. Nevertheless, I waited intently, peering at the ellipses forming on my screen as he typed.

His next message appeared. It turns out Tom wasn't gay. He was conservative.

That was the first time it fully hit me—not everyone thinks like me. In grad school, those of us who were liberal just assumed that everyone else was, too, and that our beliefs were by default correct.

Here was proof that you could know someone for years without really knowing their opinions about anything. Tom had always been supportive of my research, sending me various links to news articles on sex tech, along with words of encouragement as my studies progressed, so I assumed he was progressive. As it turns out, as much as we disagreed about a wide range of political issues, we could still find common ground and be good friends.

There is a 36:1 ratio of liberals to conservatives in academia,[2] an astounding gap that has serious implications for the knowledge being produced in the academy. I asked Jonathan Haidt about this, eager to hear his perspective. Haidt is a social psychologist at New York University, the coauthor of the *New York Times* bestselling book *The Coddling of the American Mind,* and one of the founders of Heterodox Academy, an organization of professors and graduate students committed to expanding viewpoint diversity in academia.

Haidt, who describes himself as "left-wing," told me about the lack of conservatives in academia and a proposed solution. "I had all these emails from conservative grad students and undergrads about how they were leaving because they felt really unwelcome." It got him thinking. "You know, we have to stop doing this. . . . Take all the stuff we say in social psychology—we're always talking about the benefits of diversity, well, guess what? It doesn't come from different skin color, it comes from different ideas and different values and different perspectives. So, if we take that seriously, we've got to promote viewpoint diversity."[3]

He also discussed the rising prevalence of illiberalism on campus. "The great majority of faculty are what I would call liberal-left. That means they're true liberals; they believe in freedom of speech, they believe that we should grant each other the maximum zone of freedom to live our lives. Professors are not illiberal people, but there is a small contingent of faculty and students who are illiberal, that is, who are so politicized, who see the personal as political, the academic as political—a small group." He added, "It's very few people, but all it takes is a few people to ruin your life."

This explains so much about the direction in which academia is currently headed. What's even more concerning is that this strain of illiberal-left thought doesn't actually reflect the values and beliefs of most left-leaning people. And instead of taking criticism into account, these activists continue to double down.

Moving further left across the political spectrum, however, alienates not only conservatives, but also true liberals from progressive causes. People don't appreciate having radical politics crammed down their throats at every turn. In fact, a survey from More in Common, an organization based in the United States and Europe that seeks to narrow the political divide, showed that 80 percent of Americans feel that political correctness has gone too far. It also suggested that the majority of nonwhite people actually dislike political correctness, and that it is something that highly educated, left-leaning white people have chosen to prioritize.[4]

Supporting those who are different or marginalized is something we should all strive to do, but this doesn't require denying science or disciplines like sexology. Scientific truths that may, on the surface, seem offensive or antiprogressive can in fact liberate us and help us live more fulfilling lives. They need not be seen as being incompatible with promoting human rights.

Yet the current battle consisting of far-left progressives attacking and silencing reasonable liberals, conservatives, and just about everyone else is a representation of a wider trend unfolding in every area of existence now.

Whenever I find myself in conversation with someone who isn't immersed in the culture of academia and who has been mostly following the mainstream conversation, they will say to me, "What on earth did these researchers do to deserve having their work pulled?"

First of all, no scientist deserves to have their work suppressed because some group has taken issue with it. The fact that this even happens shows how badly the process of scientific inquiry has been maligned. Only certain kinds of research are prioritized for funding, given ethical approval, and subsequently allowed to be published. For a paper that manages to survive this first set of censors despite being "controversial," once activists get wind of its publication, it will either be removed or have a correction issued while being publicly discredited.

Before we go any further, let's a take a look at exactly how we got here.

The Cost of Speaking Out

After what happened to Michael Bailey, the psychology professor at Northwestern University, most in the field decidedly went radio silent about transgender issues. They knew that expressing anything besides conciliatory agreement with trans activists was the equivalent of stomping on a hornet's nest. This is how we ended up with an open letter published by the *Guardian* and signed by fifty-four academics across disciplines as disparate as psychology, medicine, economics, computer science, and evidence-informed policy, voicing concerns about the inability to do proper research on subjects related to the transgender community.[5] The letter mentioned how harassment, campus protests, and censorship have become the norm in response to attempts at discussion and critical analysis. It also described how allegations of transphobia have been used to suppress dissenting research, and that this will likely have an impact on work that is funded and published in the future.

It's very eerie to see the activist mob calling for the head to roll of someone you know. Every sexologist will witness this at least once, if not several times, in their career. It becomes clear that if you aren't careful, there is more than a small probability that the same will happen to you. Why risk losing your job, your good name, and your privacy? It's far easier to put your (still intact) head down and wait for the tide to turn on its own.

Don't get me wrong; I get it. Speaking out is risky, especially when you're just starting out. Professors have no choice but to self-censor unless they want to risk inciting power-hungry mobbers. But that is exactly how we found ourselves in this predicament.

My sense is that most professors, and especially newly appointed ones, tell themselves to wait until they get tenure and then they'll say what they really think (not realizing that, as my mentor warned me at the start of this book, it doesn't make much of a difference in the current age). But by the time an academic reaches that milestone, they are so used to keeping their mouth shut, it's almost as though they've forgotten how to be outspoken.

Those who find themselves white-knuckling it, struggling to keep their opinions to themselves, write to me when it becomes too much to bear. I have colleagues who will distance themselves from me publicly while secretly offering words of support. Believe me, just because you don't see many people disagreeing doesn't mean they don't exist.

Some of my readers have commented that I should have stayed—how can academia be fixed if everyone keeps leaving? But if I hadn't left on my own terms, no doubt I would have been evicted shortly thereafter. Now, every time I see a scientist being berated or fired for publishing controversial findings or stating facts as plain as day, it reinforces my belief that I did the right thing.

What's scariest is that the bullies are not just Internet trolls dwelling in their parents' basement. These are professors, doctors, lawyers, and journalists who are in positions of influence, who also know that being conscious about social justice issues will help to advance their careers. They position themselves as the underdogs enforcing vigilante justice, when in actuality, they are the ones who are being oppressive.

There is nothing more cringe-worthy than watching individuals proudly proclaim their academic or scientific credentials before spewing something completing unscientific and deranged.

"I am a physician and I can tell you there is no evidence that biological sex is a relevant concept." "I have a PhD in biology, and gender is entirely a social construction." "In endocrinology, we don't fully understand the effects of testosterone."

What that says to anyone with a cursory understanding of biology is that scientists and medical professionals are willing to lie if it suits their political values. For most others who don't know the difference, they will take what these "experts" are saying at face value. This creates confusion and hampers basic science literacy. Likewise, support for these scholars isn't based on the merit of what they are saying, but on the message and the virtues they stand for.

Sex researchers have been used to dealing with right-leaning groups going out of their way to block funding of their research, portraying us as sexual deviants with no moral compass and an inability to keep our prurient proclivities to ourselves. Anytime I defend the gay community or marriage equality, a predictable slew of antigay activists climb out of the ground to tell me I'm wrong.

Now the problem is smug progressives, who seem to believe they are forward-thinking and virtuous as a result of their science denial when they are just as closed-minded and intolerant.

It is upsetting to see how hard my colleagues work, not only to do their research from a fundamental, nuts-and-bolts perspective, but also to have the field taken seriously, only to have to deal with ideologues on both sides of the political compass, making it a business to ruin their lives. In many cases, we are essentially reinventing the wheel, pretending to not know things that are commonplace knowledge, such as that men and women are different, or how many genders there are.

I understand the concern. There exists the trope of the evil scientist, experimenting on innocent patients without having to answer to anyone. There is a sordid history of scientists acting unethically, like in the case of the Tuskegee syphilis experiment, in which more than one hundred black men died because lifesaving medication, penicillin, was kept from them. Those enrolled in the study were told they were being treated for "bad blood," and the experiment was only terminated in 1972, after forty years, because whistleblowers exposed it to the public. It is commonly cited today as an example of an unethical research protocol and exactly what *not* to do as a scientist. The response from minority communities, as one might expect, has been an overall distrust of the medical system and scientific research.

Correcting Misinformation

That brings us to ways you can tell whether a scientific source is accurate and unbiased. I can't believe that this even needs to be said—that things have gotten so bad, scientific papers can't be trusted. It speaks to how prominent the practice of gaming scientific publications has become.

Keeping in mind that even rigorous and prestigious scientific journals have been swayed by fears of public complaints and retribution, here are a few ways that you can tell whether a study may be pulling the wool over your eyes: a researcher's educational background, where they got their degree, and where they are affiliated don't tell us very much these days. I'd recommend looking up their coauthors to see which university departments they are affiliated with. Physicians and hard scientists will often collaborate with faculty in nonscience departments to provide an ideological framework through which they will interpret their findings. Of course, not all scholars from the following fields are suspect, but more often than not, these predictions will ring true. Be wary of cross-disciplinary authorship with departments like philosophy, English, and education, as well as those using the word "studies" in their name, like "gender studies," "women's studies," "queer studies," "cultural studies," and the like.

Please also note—and I wish I could attach a set of flashing lights to this part—that "sexuality studies" is *not* the same thing as sexology. The former operates from a similar framework as gender studies, whereas the latter is scientific.

Also check to see if a professor's profile on the university's website contains social justice buzzwords like "inequality," "lived experience," or "the patriarchy," or calls attention to their sexual orientation, race, or status as a woman, or that they identify as a feminist. If any of their lectures or talks are publicly available, they may also serve as a valuable resource. Being sensitive to issues like inequality is important, but it isn't necessary or appropriate for scientists to make these proclamations in an academic setting. A nonideological academic knows that their identity, whether they are a "cishet white man" or otherwise, doesn't make a difference when it comes to the quality of their research.

Another clue can come from a professor's social media content, particularly that which is political. Ideological academics have been possessed by not only self-righteousness but also an aggressive tendency to shut dissenting voices up at any cost. A fair-minded academic will be willing to foster discussion about disagreement without ascribing negative motives to those critical of them, blocking the other person on social media without previous contact, or defaulting to ad hominem attacks.

More broadly, often what is reported by journalists and bloggers isn't necessarily what a study says, so I would recommend seeking out the original source, whenever possible.

Every so often, one of my colleagues outside of sex research will request lay resources they can use to counter what the elementary school system is teaching their children about gender. I've struggled to find many that were factually correct. Activist organizations have managed to infect much of the information that is available relating to both gender and biological sex. Any research studies that are not in agreement with this agenda are ignored as though they never existed. Whether it's health websites, research publications, or media articles, it really is a jungle out there. If what you are looking for is basic, foundational information, anything older than ten years old is probably safe. Anything published in the last few years is questionable. This is not to say that good resources don't exist, but I'd advocate for vetting anything you come across carefully.

Children are the most vulnerable victims of the culture war because their minds are a valuable commodity. Educators have been smart in co-opting children as part of the social justice movement, inducting them into the cult before they know any better.

I only have to peruse my local bookstore to see how parents are similarly being brainwashed. There is a multitude of books marketed at parents and therapists about exciting new developments in gender and how to raise gender-nonconforming children. Each professes that gender is an endless exploration and essentially whatever an individual wants it to be. Most concerning is that much of this material claims to be based in the latest science when it is anything but. Nevertheless, their message sounds appealing, especially when paired with a joyful array of rainbow colors in the cover art. If I didn't know anything about the science, I would likely be swayed, too.

School boards have jumped on the social justice train without a second thought. Lessons revolve around themes of systemic oppression, anti-bias, and allyship. In a New York City preschool, children are taught about "queer" rights and that people choose their gender.[6] In the Canadian province of British Columbia, one school district put up posters addressing white privilege in each of its schools.[7] Educational tools teach Canadian children struggling with their gender identity about the possibility of transitioning, including dolls[8] and a puppet who is transgender.[9]

From what I can tell, most parents don't realize that this is what their kids are being taught. The ones who have embraced it have bought the belief that it's in the next generation's best interests; others do so in order to avoid rocking the boat or being ostracized by other parents.

Some parents sense they are being hoodwinked, but because they are constantly being told that the newest science shows that gender is a social construct and fluid, they are at a loss. Considering that *adults* are finding it difficult to discern truth from fiction, one can only imagine how confusing the process will be for young children. They are dependent on parents, teachers, caregivers, and medical professionals to make decisions on their behalf based on this faulty information.

As for how parents can ensure their children aren't misinformed, every attempt at pushing back counts. The only reason this manner of thinking keeps proliferating is our silence. In the event that teachers and school officials are not amenable to suggestion or criticism, I've had parents tell me they've kept their kids home from school.

The Reality of Censorship

College campuses once encouraged critical thought and disagreement. Back in the day, if you disagreed with someone, you'd write a paper about it or have your university organize a debate. This healthy academic jousting has since been replaced with academics calling for a researcher's dismissal as icing on the cake after blacklisting and no-platforming them. Even contemplating the study of a controversial or taboo subject will immediately arouse suspicion from extreme-left academics, as they will attribute malicious intent to one's curiosity.

Professors producing meaningful work are being penalized for factors beyond their control. One of my colleagues, who is a world expert in his respective area of research, was considered as a speaker for an upcoming conference. Despite being gay and having faced discrimination in his lifetime, his name was ultimately dismissed because he was a "cisgender white man."

Some of my friends from graduate school have already faced repercussions for doing unbiased research. We finished graduate school at our respective universities around the same time, catching up every few months as they applied for professorships and interviewed for tenure-track positions, while I dodged angry mobs on social media for my various columns.

One of my friends, Kevin Hsu, who is an assistant professor of psychology at Pennsylvania State University, was a recent target of the transgender activist cabal. The controversy erupted in late 2018. At one of the major sex research conferences, he was invited to present a plenary address after winning their prestigious award for his research on gynandromorphophilia, defined earlier as sexual interest in transgender women who possess breasts and a penis.

It is, without question, a controversial area of research, and any sexologists reading this right now will probably shiver just thinking about it. To this day, Hsu has been one of few researchers who have been courageous enough to study it. The study for which he received the award showed that men with gynandromorphophilia are heterosexual (as opposed to gay, as previously believed) and that many also experience autogynephilia.[10] (See earlier chapters for a fuller discussion of these terms.)

Months before the conference took place, my colleagues in sexology were worried that something might happen. Considering trans activists' tense history with sex researchers, would one of them complain and try to tear the whole thing down?

When the day came, Hsu's presentation was relegated to a small side room, which was strange because other speakers at the conference had been given a larger conference room for their talks. Hsu nevertheless gave his presentation, only to have a trans woman, a psychologist associated with the World Professional Association for Transgender Health, repeatedly interrupt him,[11] questioning his use of particular words and taking issue with his mentioning of autogynephilia. The moderator and the audience asked her, each time, to allow Hsu to continue, as a period for questions had been allotted at the end, as is customary for research presentations. When Hsu's talk finished and the audience gave their applause, she left immediately.

After the incident, the organization's officials sent a mass email to all of its members and conference attendees. Instead of defending Hsu, as one might expect a scientific body to do, it expressed concern about his research. In what appeared to be an attempt on the part of the organization to distance itself from the award selection process, the email apologized for "more recent incidents of language and behavior that . . . made transgender persons . . . feel unwelcome, unsupported, marginalized, or attacked," with the emphasis, "We fully support you and stand with you. We are trans-allies." In response, the scientist for whom the award was named asked that his name be removed, unwilling to cow to the organization's political demands. It's worth mentioning that the statement did not offer an apology to Hsu.

After Hsu-gate, a group of researchers created an ethics document regarding appropriate language when conducting research and speaking about the transgender community. "Impact matters at least as much as intent," it read.

Hsu's experience was not the only example in recent memory of activist politics interfering with sex research. Another researcher had his study halted by his university's institutional review board. Despite having the necessary funding and approval from his department, the ethics committee was concerned about some aspect of the study, but would not give him a clear answer as to what it was. This is not the norm as part of the ethical review process. Usually a researcher is given very specific feedback as to what needs to change in their protocol in order for a study to be granted approval.

The researcher suspected it was because his study involved the LGBT+ community, which left him in the undesirable position of having to disclose his sexual orientation as a gay man—in front of a committee of more than a dozen people—in hopes that it might assuage their fears and alter their decision.

It did not. In addition to having his research denied, what was particularly distasteful to me, and what made me the maddest for him, is that no one should have to reveal something as personal as their sexual orientation as a justification for their work.

Perhaps there was another reason why the study was shut down. But if so, why not give him a clear answer? If the ethics committee had concerns about the implications for this marginalized community, how is it empowering to deny one of its members the ability to better understand it and educate the public? A previous research paper has suggested that institutional review boards may be concerned about research coercion in this vulnerable population. Because the LGBT+ community has experienced ethical transgressions in the past, ethics committees may be overprotective of the community, creating unnecessary hoops for researchers to jump through, and unfortunately, leading to less research being done.[12]

We can't forget what happened to Lisa Littman's study on rapid-onset gender dysphoria (see Chapter 5). There was also the censorship twice over of Theodore Hill, a professor emeritus of mathematics at the Georgia Institute of Technology, after writing about the "greater male variability hypothesis," which proposed that a greater number of men exist at the extreme high and low ends of intelligence. After his paper was accepted for publication and rescinded the second time, Hill was told this was due not to its scientific approach but its political consequences.[13]

In a similar twist of fate, James Caspian, a former master's student at Bath Spa University, sought to study detransitioners. His proposed thesis was rejected by the university's ethics committee for being "potentially 'politically incorrect.' "[14]

In late 2019, another scientific journal published a study on the cognitive mechanisms of gender dysphoria. This particular journal is the official open-access journal published by one of the largest professional organizations for neuroscientists globally.

The study suggested that gender dysphoria is the result of differences in neural networks pertaining to sensory perception and ownership of one's body parts, and has less to do with sexual dimorphism in the brain. This is in stark opposition to the current dominant narrative that gender dysphoria is due to having the wrong brain sex.

The paper proposed that treatment for gender dysphoria should focus on lessening feelings of distress, which suggested that gender dysphoria may be a condition that is transient and capable of being resolved without transitioning. Every expert under the sun admonished this idea, and following a post-publication review, the paper was retracted.

Although such an extrapolation may not apply to everyone with gender dysphoria, it has relevance. The vast rates of detransitioning we're witnessing suggest that transitioning isn't always the best solution. The research and clinical communities should be open to alternative perspectives that are science-based, particularly in the context of scholarly debate.

These small wins send a very clear message to ideologues: science can be rewritten if you throw a loud enough temper tantrum.

Returning to the research talk I attended, in which I found myself mouth agape with silent horror, just when I thought things couldn't get worse, the presenter pulled up his next slide, which consisted of two cartoon figures explaining how there were other genders besides male and female, and that sexual orientation, gender identity, gender expression, and anatomy were unrelated. Using a laser pointer, he began explaining somewhat shakily how a person could identify along any of these continua however they wanted, and knowing one such piece of information about an individual didn't translate to knowing anything else about them.

He doesn't even believe what he's saying, it seemed to me. *He's only saying this because he has to.*

A room full of impressionable, young graduate students were, at the same time, learning one of two things: even a scientist from a highly respected institution must yield to social justice doctrines, and it was possible within a matter of seconds for basic truths, uncontested in the field, to vanish into thin air, as though they never existed to begin with.

How demoralizing an existence must it be. But what is the alternative when a misstep will cost you the career you worked so hard for? Most disturbingly, these teaching materials were clearly marketed for educators to use in children's classrooms.

Sexology isn't the only field going down in flames due to activist meddling. There have been casualties across a variety of other academic disciplines, including intelligence research, interplanetary exploration, and something as seemingly benign as treatment for chronic fatigue syndrome.[15] Researchers will change their areas of study because the online abuse and penalties to their personal reputation are too great. It's not an issue of one or two specific questions being flagged as out of bounds, but rather, entire areas of inquiry.

The soul-sucking, joyless nature of social justice has seeped into lifestyle and hobbyist communities, too, including—of all places—knitting,[16] wherein allegations of racism and colonization in the online community led to targeted harassment of several prominent knitters. In every such case, there is some sort of purity test that an individual fails, and as a result, they are sicced upon by the mob.

In the pursuit of scientific answers, who ends up winning? Marginalized communities? What good does it do if all of the impartial experts in the field are too terrified to engage with you? In the end, the only researchers willing to study these populations will be those confident that their results won't upset anyone. But what they will be doing at that point will no longer be science.

CONCLUSION

THE END OF ACADEMIC FREEDOM

I remain close with a number of my sexological colleagues, and at least once a week, I get an email highlighting yet another level of absurdity that my former discipline has reached. My response to these updates is usually shock at how grotesque it's become, followed by laughter in utter disbelief. Better to laugh than cry, right?

As you now know, social justice has managed to seep beyond the confines of the academic community into the real world. Near the end of my doctoral studies, I received a very kind invitation from an organization asking if I was interested in speaking at an upcoming event. It would include experts from a variety of disciplines, including law and

HR, aimed at educating private corporations on issues relating to gender in the workplace.

My talk would be about the science of gender, covering whatever aspects I wanted. Fantastic! I was excited to take my technical skills and use them in an applied setting.

As I began making my slides, I soon saw how many of the topics I wanted to cover were murkier than I had realized. Maybe this was a bad idea.

That's okay, I thought, maintaining my optimism. *I'll find a way to make this work.*

As I wrapped up my preparation a few days later, my positivity had been hardened by what lay in front of me. I realized that I couldn't lie, which was surely an indication of what was to come.

The event took place on the top floor of a high-rise, in a brightly painted conference room with gold-flecked pens on every table. I gave my talk, which was straightforward enough. At the end, the organizer opened up the floor for questions. One of the attendees raised her hand. She had a frown on her face. The gist of our conversation went like this:

"Gender is a social construct," she said.

"No, it isn't," I said.

"There are new studies showing no differences between men and women in the brain."

"Those studies are wrong."

The room suddenly got very quiet. One male employee, who had been poking around on his phone for the entirety of my talk, stopped what he was doing to look up at me.

"Next question," I said.

A few people asked me to elaborate on some of the animal studies I had mentioned. Then another woman, in a cream-colored pantsuit, bravely put up her hand. Her organization was doing away with the concept of gender and had announced it would be implementing all-gender bathrooms.

"If there is no such thing as gender," she began, "how do we differentiate between men and women?"

Confused, I asked her to clarify.

"There are some things that are usually associated with being female," she said. "But if we can't use the word 'female,' how do we refer to them?"

I didn't know what to tell her. What I wanted to say was, "You refer to them as female, and tell anyone suggesting otherwise to take a hike." But I inadvertently caught a glance at the organizer, who had this pleading look in her eyes, and not wanting to alienate the other half of the room that wasn't yet angry at me, I remained speechless.

We took a break for lunch. As any graduate student can tell you, the biggest incentive for going anywhere in life is free food. As happy as I was to be getting a paycheck from the event, I was just as delighted to be embarking on an all-you-can-eat feast. It didn't matter that I had earned the title of "hatemonger" that day. I was going to enjoy the catering.

The most curious thing happened next. As I collected my lunch, some members of the audience would pretend they were interested in helping themselves to whichever catering tray I was shoveling food onto my plate from, when really, it was to thank me privately.

"Thank you for saying that," they would say, before ducking away quickly.

After lunch, things only got worse. The rest of the speakers that day were clearly high on social justice and their interpretation of policies relating to law and medicine were heavily steeped not in science or fact, but activist thinking. The final session of the day consisted of us reading sample vignettes aloud to one another in an attempt to inspire greater authenticity around expressing gender in our lives.

I was reminded of one of my friends, who bought a house a few years earlier, with the goal of renovating it. Within months, he discovered it had a full-blown ant infestation. Anytime I was there paying him a visit, we would spend a good hour stamping out his new little friends. He also showed me, with a weird mix of revulsion and pride, that he had uncovered an entire colony tucked behind the insulation. What appeared, at first glance, to be a healthy wooden frame was in fact housing multiple fervid nests of the queen's eggs.

Gender ideology reminds me a lot of having an insect problem. It is insidious, popping up in clusters when you least expect it. When it does, you can't ever get away from it. No matter how many times you try to exterminate the festering, there is another manifestation, lying in wait. These employees would return to their respective companies in the day following the training, and these ideas would continue to permeate.

Sex research conferences, once fertile (I couldn't help myself) grounds for intellectual inquiry, have already ceded much territory. Conferences are now overrun with preferred pronouns on name cards, anti-oppression workshops, and initiatives catering to "queer" people (never mind that the word "queer" is still considered by many gay people to be a homophobic term of abuse, as we saw in Chapter 3). Plenary talks are predictably infused with feminism. Feminist research methods. Intersectionality in sexual health. Decolonizing sexuality. And just when you thought you were done for the day, conference organizers ensured that social justice activities were planned as evening events.

The trend is also evident in academic research job postings, which seem to place a greater emphasis on whether a potential candidate belongs to the population being studied than their skills as a researcher. Other job postings cite a specific quota that needs to be filled, such as research assistants who identify as nonbinary. It's almost to the point that a commitment to equity, inclusion, and ticking identity politics boxes is more important than one's productivity as a scientist.

I worry for the next generation of sex researchers. Those who are more focused on science than activism will self-select out of the field because they recognize how toxic it is becoming. The older generation of scientists, who preferred to focus on the content and quality of their work instead of its political reach, is retiring, only to be replaced by social justice believers. Even after a person has left the field, they aren't fully free. To this day, I still get lumped into strange conspiracy theories about how sexologists are plotting to eradicate certain communities.

All hope is not lost, however. I do think science will eventually prevail; it may simply take some time before we get there. As I said at the start of this book, the truth can be suppressed, but it will always come out. In the meantime, those of us who have been banished to the wrong side of history have learned a thing or two about maneuvering amid the hysteria (wink, wink). After seeing a number of writers have their books pulled from publication at the hands of activists, I kept very quiet throughout the process of writing this one.

As for those who are still in the academic trenches fighting the good fight, I say, keep going. Don't apologize for whatever it is you find in your research. In turn, journal editors and reviewers should be firm about which studies are acceptable for publication and which do not deserve to see daylight, instead of caving to activists' complaints and backpedaling once research findings have been published.

To play devil's advocate for a moment, must sexology and social justice really be at each other's throats? Are the two really incompatible, or is there a compromise? Can there exist a middle ground upon which both can be integrated? I do believe scientists owe it to the public to be as transparent and accountable as possible regarding the research they are conducting, and to take feedback from relevant communities into consideration. This would help to neutralize a lot of the animosity and mistrust directed at them from groups that have been discriminated against and exploited in the past. In return, activists should respect the scientific process and have confidence that those who are unethical will be called out by other scientists and disciplined by their institutions.

Experiencing unease, or disliking a particular study's findings, is not a justification for shutting down the conversation around the issue altogether. As is commonly the case with censorship, attempting to smother dissenting views only makes these views more appealing to impartial onlookers.

For those who say that hateful ideas shouldn't be entertained and that debating someone you disagree with legitimizes their position, muzzling the debate doesn't make them go away. Controversial research, in itself, doesn't pose a threat. Instead of attacking and punishing researchers, we should be combating those who misuse research findings to uphold their prejudicial views.

Of all places, universities should be the strongest proponents of the First Amendment, not enthusiastic supporters of stifling speech and open discussion. Instead, the institutions themselves have been happy to capitulate to the demands of those who have no respect for the academic process or intellectual discourse.

I hope that this book has helped to arm you with facts for the battles you face in your own lives. In time, I don't doubt that the pendulum will swing back to the center before continuing on toward the opposite extreme. The propensity for science denial will always be there, because the truth about who we are is uncomfortable. Opinions are too quickly made based on which side you stand on and what the right answer is for your side. Sometimes we have to take a breath, pull ourselves back, and ask why we believe the things that we do.

ACKNOWLEDGMENTS

This book would not have been possible without the support of my readers, especially those of you who have been with me since day one with the publication of my very first column. Deciding to leave academia to become a journalist was such a precarious decision, and I could not have done it without you.

A heartfelt thank-you to Natasha Simons, for seeing the potential for this book after coming across a serendipitous interview of me on the Internet. You gave me the freedom to say what I needed to say and made my childhood dream come true. Steve Troha, I am grateful for your representation, all of your advice and insight, and your wicked sense of humor.

Many thanks to the Simon & Schuster family, especially Maggie Loughran, Jennifer Long, Jennifer Bergstrom, Abby Zidle, and Al Madocs. Thank you to Jennifer Weidman for keeping me out of trouble; Lauren Truskowski and Jennifer Robinson, for helping me get my message out; John Vairo and Lisa Litwack, for designing the perfect cover; and Tom Pitoniak, for his impeccable copyediting.

Thank you for your support and encouragement; it means the world to me: Margaret Wente, Natasha Hassan, Joe Rogan, Bill Maher and everyone at *Real Time*, Bari Weiss, Dan Savage, Steven Pinker, Ben Shapiro, Eric Weinstein, Simon Baron-Cohen, Heather Mac Donald, Christina Hoff Sommers, Danielle Crittenden, David Frum, Cooper Hefner, the *Playboy* team, Michael Shermer, David Buss, Bret Weinstein, Nick Gillespie, Gerard Baker, Juliet Lapidos, Greg Gutfeld, Dave Rubin, Glenn Beck, Sara Gonzales, Kevin Ryan, Samantha Sullivan, Luke Thomas, Mish Barber-Way, Art Tavana, Lee Jussim, and Barrett Wilson.

A special thank-you to Lou Perez, We The Internet, and the Moving Picture Institute, for sharing my love of dry humor and collaborating with me so early on.

I am indebted to your expertise, guidance, and kindness: Buck Angel, Susan Bradley, Ray Blanchard, Michael Bailey, Kevin Hsu, Erik Wibowo, Anthony Bogaert, Jonathan Haidt, Simon LeVay, Thomas Steensma, Will Malone, Michael Laidlaw, Meng-Chuan Lai, Kenneth Zucker, Peggy Cadet, JP de Ruiter, Lawrence Williams, Ryan Bone, Andreas Kalogiannides, and the detransitioners who shared their stories with me.

To my friends and family—I wish I could name you here, but we all know what would happen if I did. I love you and thank you for believing in me.

NOTES

Introduction: The Battle Against Biology

1 French, D. (2015, September 2). Common sense, part II: Not every sex researcher thinks young kids should "transition." *National Review.* Retrieved from https://www.nationalreview.com/corner/common-sense-part-ii-not-every-sex-researcher-thinks-young-kids-should-transition/.

Myth #1: Biological Sex Is a Spectrum

1 Bramble, M. S., Roach, L., Lipson, A., Vashist, N., Eskin, A., Ngun, T., . . . Vilain, E. (2016). Sex-specific effects of testosterone on the sexually dimorphic transcriptome and epigenome of embryonic neural stem/progenitor cells. *Nature's Scientific Reports, 6,* 1–13.

2 For an example, see 5 misconceptions about sex and gender. (2019, March 29). [Video file]. Retrieved from https://www.youtube.com/watch?v=2S0e-i117vY.

3 Arboleda, V. A., Sandberg, D. E., & Vilain, E. (2014). DSDs: genetics, underlying pathologies and psychosexual differentiation. *Nature Reviews Endocrinology, 10,* 603–615.

4 Jones, T., & Leonard, W. (2019). *Health and wellbeing of people with intersex variations: Information and resource paper* (Intersex Expert Advisory Group). Melbourne, Australia: Victorian Department of Health and Human Services.

5 Kiprop, V. (2019, July 18). The world's population by eye color percentages. [Weblog]. Retrieved from https://www.worldatlas.com/articles/which-eye-color-is-the-most-common-in-the-world.html.

6 Cunningham, A. L., Jones, C. P., Ansell, J., & Barry, J. D. (2010). Red for danger? The effects of red hair in surgical practice. *British Medical Journal, 341,* 1304–1305.

7 Transgender, intersex, and gender non-conforming people #WontBeErased by pseudoscience. (2018, October 26). Retrieved from https://not-binary.org/statement/.

8 According to the *New York Times*, the memo described gender as being "a biological, immutable condition determined by genitalia at birth" that should be determined "on a biological basis that is clear, grounded in science, objective and administrable." See Green, E. L., Benner, K., & Pear, R. (2018, October 21). "Transgender" could be defined out of existence under Trump administration. *New York Times.* Retrieved from https://www.nytimes.com/2018/10/21/us/politics/transgender-trump-administration-sex-definition.html.

9 The *Guardian* reported that the White House was attempting "to define gender as unchangeable from birth." See Holpuch, A., & Walters, J. (2018, October 23). "It still scares me": Panic simmers below Trump trans policy protests. *Guardian.* Retrieved from https://www.theguardian.com/society/2018/oct/23/trump-transgender-policy-protests-panic-fear.

10 For additional information, see Soh, D. W. (2016, August 30). Let's make sure Ontario's sex-ed curriculum is here to stay. *Globe and Mail.* Retrieved from http://www.theglobeandmail.com/opinion/lets-make-sure-ontarios-sex-ed-curriculum-is-here-to-stay/article31605288/, and Soh, D. W. (2018, July 15). Ontario's sex-ed backlash isn't about children's safety. *Globe and Mail.* Retrieved from https://www.theglobeandmail.com/opinion/article-ontarios-sex-ed-backlash-isnt-about-childrens-safety/.

11 Crowe, J. (2019, May 9). California sex ed guidelines encourage teachers to explain gender identity to kindergartners. *National Review.* Retrieved from https://www.nationalreview.com/news/california-sex-ed-guidelines-encourage-teachers-to-explain-gender-identity-to-kindergartners/.

Myth #2: Gender Is a Social Construct

1 Del Giudice, M., Lippa, R. A., Puts, D. A., Bailey, D. H., Bailey, J. M., & Schmitt, D. P. (2016). Joel et al.'s method systematically fails to detect large, consistent sex differences. *Proceedings of the National Academy of Sciences of the United States of America, 113,* E1965.

For the other three studies mentioned, see Chekroud, A. M., Ward, E. J., Rosenberg, M. D., & Holmes, A. J. (2016). Patterns in the human brain mosaic discriminate males from females. *Proceedings of the National Academy of Sciences of the United States of America, 113,* E1968; Glezerman, M. (2016). Yes, there is a female and a male brain: Morphology versus functionality. *Proceedings of the National Academy of Sciences of the United States of America, 113,* E1971; Rosenblatt, J. D. (2016). Multivariate revisit to "sex beyond the genitalia." *Proceedings of the National Academy of Sciences of the United States of America, 113,* E1966–E1967.

2 Tan, A., Ma, W., Vira, A., Marwha, D., & Eliot, L. (2016). The human hippocampus is not sexually-dimorphic: Meta-analysis of structural MRI volumes. *Neuroimage, 124,* 350–366.

3 Marwha, D., Halari, M., & Eliot, L. (2017). Meta-analysis reveals a lack of sexual dimorphism in human amygdala volume. *Neuroimage, 147,* 282–294.

4 For a classic example, see Halari, R., Sharma, T., Hines, M., Andrew, C., Simmons, A., & Veena, K. (2006). Comparable fMRI activity with differential behavioural performance on mental rotation and overt verbal fluency tasks in healthy men and women. *Experimental Brain Research, 169,* 1–14.

5 For an example, see Stoléru, S., Fonteille, V., Cornélis, C., Joyal, C., & Moulier, V. (2012). Functional neuroimaging studies of sexual arousal and orgasm in healthy men and women: A review and meta-analysis. *Neuroscience and Biobehavioural Reviews, 36,* 1481–1509.

6 Ritchie, S. J., Cox, S. R., Shen, X., Lombardo, M. V., Reus, L. M., Alloza, C., . . . Deary, I. J. (2018). Sex differences in the adult human brain: Evidence from 5216 UK Biobank participants. *Cerebral Cortex, 28,* 2959–2975.

7 Lotze, M., Domin, M., Gerlach, F. H., Gaser, C., Lueders, E., Schmidt, C. O., & Neumann, N. (2019). Novel findings from 2,838 adult brains on sex differences in gray matter brain volume. *Nature's Scientific Reports, 9,* 1–7.

8 Ingalhalikar, M., Smith, A., Parker, D., Satterthwaite, T. D., Elliott, M. A., Ruparel, K., . . . Verma, R. (2013). Sex differences in the structural connectome of the human brain. *Proceedings of the National Academy of Sciences of the United States of America, 111,* 823–828.

9 Cahill, L. (2014). Equal ≠ the same: Sex differences in the human brain. *Cerebrum, 5,* 1–19.

10 Bennett, C.M., Baird, A. A., Miller, M. B., & Wolford G. L. (2010). Neural correlates of interspecies perspective taking in the post-mortem Atlantic salmon: An argument for proper multiple comparisons correction. *Journal of Serendipitous and Unexpected Results, 1,* 1–5.

11 Loomes, R., Hull, L., & Mandy, W. P. L. (2017). What is the male-to-female ratio in autism spectrum disorder? A systematic review and meta-analysis. *Journal of the American Academy of Child & Adolescent Psychiatry, 56,* 466–474.

12 For an example, see 2018 NCWIT Summit—Plenary II, pink brain, blue brain: What's the real story when it comes to "gender differences" presented by Lise Eliot. (2018). Retrieved from https://www.ncwit.org/video/2018-ncwit-summit-plenary-ii-pink-brain-blue-brain-whats-real-story-when-it-comes-gender.

13 Conger, K. (2017, August 5). Exclusive: Here's the full 10-page anti-diversity screed circulating internally at Google. *Gizmodo.* Retrieved from https://gizmodo.com/exclusive-heres-the-full-10-page-anti-diversity-screed-1797564320/amp.

14 Nicas, J. (2017, August 6). Google's new diversity chief criticizes employee's memo. *Wall Street Journal.* Retrieved from https://www.wsj.com/articles/googles-new-diversity-chief-criticizes-employees-memo-1502037022.

15 Ibid., note 13.

16 Coren, M. (2017, August 19). James Damore is proving the alt-right playbook can work in Silicon Valley. *Quartz.* Retrieved from https://qz.com/1055466/the-alt-right-has-an-all-new-formula-for-undermining-silicon-valley/.

17 Dockray, H. (2017, August 11). A "conversation" with Google's recently fired Tech-Bro-in-Chief. *Mashable.* Retrieved from https://mashable.com/2017/08/11/james-damore-google-rebuttal/.

18 Jaschik, S. (2005, February 18). What Larry Summers said. *Inside Higher Ed.* Retrieved from https://www.insidehighered.com/news/2005/02/18/what-larry-summers-said.

19 Ingram, D., & Palli, I. (2017, August 7). Google fires employee behind anti-diversity memo. *Reuters.* Retrieved from https://www.reuters.com/article/us-google-diversity/google-fires-employee-behind-anti-diversity-memo-idUSKBN1AO088.

20 Hopkins, A. (2019, June 10). Judge rules lawsuit accusing Google of bias against conservatives can proceed. *Fox News.* Retrieved from https://www.foxnews.com/tech/google-conservative-bias-case-james-damore.

21 Soh, D. W. (2017, August 8). No, the Google manifesto isn't sexist or anti-diversity. It's science. *Globe and Mail.* Retrieved from https://www.theglobeandmail.com/opinion/no-the-google-manifesto-isnt-sexist-or-anti-diversity-its-science/article35903359/.

22 The Google memo: Four scientists respond (2017, August 7). *Quillette.* Retrieved from https://quillette.com/2017/08/07/google-memo-four-scientists-respond/.

23 Kallak, T. K., Hellgren, C., Skalkidou, A., Sandelin-Francke, L., Ubhayasekhera, K., Bergquist, J., . . . Sundström Poromaa, I. (2017). Maternal and female fetal testosterone levels are associated with maternal age and gestational weight gain. *European Journal of Endocrinology, 177,* 379–388.

24 Hines, M. (2006). Prenatal testosterone and gender-related behaviour. *European Journal of Endrocrinology, 15S,* S115–S121.

25 Schmitt, D. P. (2015). The evolution of culturally-variable sex differences: Men and women are not always different, but when they are . . . it appears not to result from patriarchy or sex role socialization. In T. K. Shackelford & R. D. Hansen, (Eds.), *The evolution of sexuality* (pp. 221–256). Switzerland: Springer International Publishing.

26 Episode 576: When Women Stopped Coding (2016, July 22). *Planet Money.* Podcast retrieved from https://www.npr.org/sections/money/2016/07/22/487069271/episode-576-when-women-stopped-coding.

27 Jussim, L. (2017, July 20). Why brilliant girls tend to favor non-STEM careers. *Psychology Today.* Retrieved from https://www.psychologytoday.com/ca/blog/rabble-rouser/201707/why-brilliant-girls-tend-favor-non-stem-careers.

28 Lewis, P. (2017, November 17). "I see things differently": James Damore on his autism and the Google memo. *Guardian.* Retrieved from https://www

.theguardian.com/technology/2017/nov/16/james-damore-google-memo-interview-autism-regrets.

29 National Academies of Sciences, Engineering, and Medicine (2018). *Sexual harassment of women: Climate, culture, and consequences in academic sciences, engineering, and medicine.* Washington, DC: National Academies Press.

30 Williams, W. M., & Ceci, S. J. (2015). National hiring experiments reveal 2:1 faculty preference for women on STEM tenure track. *Proceedings of the National Academy of Sciences of the United States of America, 112,* 5360–5365.

31 Ceci, S. J., Ginther, D. K., Kahn, S., & Williams, W. M. (2014). Women in academic science: A changing landscape. *Psychological Science in the Public Interest, 15,* 75–141.

Myth #3: There Are More than Two Genders

1 Lane, R. (2019). Developing inclusive primary care for trans, gender-diverse and nonbinary people. *Canadian Medical Association Journal, 191,* E61–E62.

2 GLAAD (2017). Accelerating acceptance: A Harris Poll survey of Americans' acceptance of LGBTQ people. Retrieved from https://www.glaad.org/files/aa/2017_GLAAD_Accelerating_Acceptance.pdf.

3 Pew Research Center (2019, January 14). Generation Z looks a lot like millennials on key social and political issues. Retrieved from https://www.pewsocialtrends.org/2019/01/17/generation-z-looks-a-lot-like-millennials-on-key-social-and-political-issues/psdt_1-17-19_generations-15/.

4 Intersex Society of North America (2008). Why doesn't ISNA want to eradicate gender? Retrieved from https://isna.org/faq/not_eradicating_gender/.

5 Shupe, J. (2019, March 10). I was America's first "nonbinary" person. It was all a sham. *Daily Signal.* Retrieved from https://www.dailysignal.com/2019/03/10/i-was-americas-first-non-binary-person-it-was-all-a-sham/.

6 For an example, see Macrae, C. N., & Bodenhausen, G. V. (2000). Social cognition: Thinking critically about others. *Annual Review of Psychology, 51,* 93–120.

7 For example, see Kaul, C., Rees, G., & Ishai, A. (2011). The gender of face stimuli is represented in multiple regions in the human brain. *Frontiers in Human Neuroscience, 4,* 1–12.

8 Wong, Y. J., Ho, M.-H. R., Wang, S.-Y., & Miller, I. S. K. (2017). Meta-analyses of the relationship between conformity to masculine norms and mental health–related outcomes. *Journal of Counseling Psychology, 64,* 80–93.

9 Liszewski, W., Peebles, J, K., Yeung, H., & Arron, S. (2018). Persons of nonbinary gender—awareness, visibility, and health disparities. *New England Journal of Medicine, 379,* 2391–2393.

Myth #4: Sexual Orientation and Gender Identity Are Unrelated

1 For example, see Klein, F., Sepekoff, B., & Wolf, T. J. (1985). Sexual orientation: A multi-variable dynamic process. *Journal of Homosexuality, 11,* 35–49.

2 Blanchard, R. (2004). Quantitative and theoretical analyses of the relation between older brothers and homosexuality in men. *Journal of Theoretical Biology, 230,* 173–187.

3 LeVay, S. (1991). A difference in hypothalamic structure between heterosexual and homosexual men. *Science, 253,* 1034–1037.

4 Safron, A., Klimaj, V., Sylva, D., Rosenthal, A. M., Li, M., Walter, M., & Bailey, J. M. (2018). Neural correlates of sexual orientation in heterosexual, bisexual, and homosexual women. *Nature's Scientific Reports, 8,* 1–14.

5 Safron, A., Sylva, D., Klimaj, V., Rosenthal, A. M., Li, M., Walter, M., & Bailey, J. M. (2017). Neural correlates of sexual orientation in heterosexual, bisexual, and homosexual men. *Nature's Scientific Reports, 7,* 1–15.

6 Lewis, G. B. (2009). Does believing homosexuality is innate increase support for gay rights? *Policy Studies Journal, 37,* 669–693.

7 Bailey, J. M., & Zucker, K. J. (1995). Childhood sex-typed behavior and sexual orientation: A conceptual analysis and quantitative review. *Developmental Psychology, 31,* 43–55.

8 Green, R. (1987). *The "sissy boy syndrome" and the development of homosexuality.* New Haven, CT: Yale University Press.

9 Henley, C. L., Nunez, A. A., & Clemens, L. G. (2011). Hormones of choice: The neuroendocrinology of partner preference in animals. *Frontiers in Neuroendocrinology, 32,* 146–154.

10 Hines, M. (2011). Sex-related variation in human behavior and the brain. *Trends in Cognitive Science, 14,* 448–456.

11 Hines, M. (2011). Prenatal endocrine influences on sexual orientation and on sexually differentiated childhood behavior. *Frontiers in Neuroendocrinology, 32,* 170–182.

12 Bakker, J. (2018). Brain structure and function in gender dysphoria. *Endrocrine Abstracts, 56,* S30.3.

13 VanderLaan, D. P., Lobaugh, N. J., Chakravarty, M. M., Patel, R., Chavez, S., Stojanovski, S. O., . . . Zucker, K. J. (2015). The neurohormonal hypothesis of gender dysphoria: Preliminary evidence of cortical surface area differences in adolescent natal females. Poster session presented at the International Academy of Sex Research Annual Conference, Toronto, ON.

14 Rametti, G., Carrillo, B., Gómez-Gil, E., Junque, C., Zubiarre-Elorza, L., Segovia, S., . . . Guillamon, A. (2011). The microstructure of white matter in male to female transsexuals before cross-sex hormonal treatment: A DTI study. *Journal of Psychiatric Research, 45,* 949–954.

15 Rametti, G., Carrillo, B., Gómez-Gil, E., Junque, C., Segovia, S., Gomez, Á,

& Guillamon, A. (2011). White matter microstructure in female to male transsexuals before cross-sex hormonal treatment. A diffusion tensor imaging study. *Journal of Psychiatric Research, 45,* 199–204.

16 Campo, J., Nijman, H., Merckelbach, H., & Evers, C. (2003). Psychiatric comorbidity of gender identity disorders: A survey among Dutch psychiatrists. *American Journal of Psychiatry, 160,* 1332–1336.

17 Murad, H. M., Elamin, M. B., Garcia, M. Z., Mullan, R. J., Murad, A., Erwin, P. J., & Montori, V. M. (2010). Hormonal therapy and sex reassignment: A systematic review and meta-analysis of quality of life and psychosocial outcomes. *Clinical Endocrinology, 72,* 214–231.

18 Dhejne, C., Lichtenstein, P., Boman, M., Johansson, A. L. V., Langstrom, N., Landén, M. (2011). Long-term follow-up of transsexual persons undergoing sex reassignment surgery: Cohort study in Sweden. *Proceedings of National Academy of Sciences of the United States of America, 6,* 1-8.

19 Blanchard, R. (2005). Early history of the concept of autogynephilia. *Archives of Sexual Behavior, 34,* 439–446.

20 Lawrence, A. A., & Bailey, J. M. (2008). Transsexual groups in Veale et al. (2008) are "autogynephilic" and "even more autogynephilic." *Archives of Sexual Behavior, 38,* 173–175.

21 Hsu, K. J., Rosenthal, A. M., & Bailey, J. M. (2015). The psychometric structure of items assessing autogynephilia. *Archives of Sexual Behavior, 44,* 1301–1312.

22 Blanchard, R. (2005). Early history of the concept of autogynephilia. *Archives of Sexual Behavior, 34,* 439–446.

23 Blanchard, R. (2008). Deconstructing the feminine essence narrative. *Archives of Sexual Behavior, 37,* 434–438.

24 Savic, I., & Arver, S. (2011). Sex dimorphism of the brain in male-to-female transsexuals. *Cerebral Cortex, 21,* 2525–2533.

25 Moser, C. (2009). Autogynephilia in women. *Journal of Homosexuality, 56,* 539–547.

26 Dreger, A. D. (2008). The controversy surrounding *The Man Who Would Be Queen*: A case history of the politics of science, identity, and sex in the Internet age. *Archives of Sexual Behavior, 37,* 366–421.

27 Zucker, K. J., Bradley, S. J., Owen-Anderson, A., Kibblewhite, S. J., Wood, H., Singh, D., & Choi, K. (2012). Demographics, behavior problems, and psychosexual chracteristics of adolescents with gender identity disorder or transvestic fetishism. *Journal of Sex and Marital Therapy, 38,* 151–189.

28 Perry, L. (2019, November 6). What is autogynephilia? An interview with Dr. Ray Blanchard. *Quillette.* Retrieved from https://quillette.com/2019/11/06/what-is-autogynephilia-an-interview-with-dr-ray-blanchard/.

29 Ibid., note 28.

30 Wood, H., Sasaki, S., Bradley, S. J., Singh, D., Fantus, S., Owen-Anderson, W., . . . Zucker, K. (2013). Patterns of referral to a gender identity service for children and adolescents (1976–2011): Age, sex ratio, and sexual orientation. *Journal of Sex and Marital Therapy, 39,* 1–6.

31 Chivers, M. L., Roy, C., Grimbos, T., Cantor, J. M., & Seto, M. C. (2014). Specificity of sexual arousal for sexual activities in men and women with conventional and masochistic sexual interests. *Archives of Sexual Behavior, 43,* 931–940.

Myth #5: Children with Gender Dysphoria Should Transition

1 All eleven studies are as listed:

Davenport, C. W. (1986). A follow-up study of 10 feminine boys. *Archives of Sexual Behavior, 15,* 511–517.

Drummond, K. D., Bradley, S. J., Badali-Peterson, M., & Zucker, K. J. (2008). A follow-up study of girls with gender identity disorder. *Developmental Psychology, 44,* 34–45.

Green, R. (1987). *The "sissy boy syndrome" and the development of homosexuality.* New Haven, CT: Yale University Press.

Kosky, R. J. (1987). Gender-disordered children: Does inpatient treatment help? *Medical Journal of Australia, 146,* 565–569.

Lebovitz, P. S. (1972). Feminine behavior in boys: Aspects of its outcome. *American Journal of Psychiatry, 128,* 1283–1289.

Money, J., & Russo, A. J. (1979). Homosexual outcome of discordant gender identity/role: Longitudinal follow-up. *Journal of Pediatric Psychology, 4,* 29–41.

Singh, D. (2012). A follow-up study of boys with gender identity disorder. Unpublished doctoral dissertation, University of Toronto.

Steensma, T. D., McGuire, J. K., Kreukels, B. P. C., Beekman, A. J., & Cohen-Kettenis, P. T. (2013). Factors associated with desistence and persistence of childhood gender dysphoria: A quantitative follow-up study. *Journal of the American Academy of Child and Adolescent Psychiatry, 52,* 582–590.

Wallien, M. S. C., & Cohen-Kettenis, P. T. (2008). Psychosexual outcome of gender-dysphoric children. *Journal of the American Academy of Child and Adolescent Psychiatry, 47,* 1413–1423.

Zuger, B. (1978). Effeminate behavior present in boys from childhood: Ten additional years of follow-up. *Comprehensive Psychiatry, 19,* 363–369.

Zuger, B. (1984). Early effeminate behavior in boys: Outcome and significance for homosexuality. *Journal of Nervous and Mental Disease, 172,* 90–97.

2 Steensma, T. D., Biemond, R., de Boer, F., & Cohen-Kettenis, P. T. (2011). Desisting and persisting gender dysphoria after childhood: A qualitative follow-up study. *Clinical Child Psychology and Psychiatry, 16,* 499–516.

3 Ruble, D. N., Martin, C. L., & Berenbaum, S. A. (2006). Gender development. In W. Damon & R. M. Lerner (series eds.) and N. Eisenberg (vol. ed.), *Handbook of Child Psychology* (6th ed.). *Vol. 3: Social, Emotional, and Personality Development* (pp. 858–932). New York: Wiley.

4 Drummond, K. D., Bradley, S. J., Badali-Peterson, M., & Zucker, K. J. (2008). A follow-up study of girls with gender identity disorder. *Developmental Psychology, 44,* 34–45.

5 Ibid., note 1.

6 Reid, S. (2020, January 24). "Why did the NHS let me change sex?" Star witness in court battle against clinic that fast-tracked her gender swap aged 16 reveals what happened when she made a cry for help. *Daily Mail.* Retrieved from https://www.dailymail.co.uk/news/article-7926675/Witness-court-battle-against-gender-clinic-reveals-happened-cry-help.html.

7 Steensma, T. D., McGuire, J. K., Kreukels, B. P., Beekman, A. J., & Cohen-Kettenis, P. T. (2013). Factors associated with desistence and persistence of childhood gender dysphoria: A quantitative follow-up study. *Journal of the American Academy of Child and Adolescent Psychiatry, 52,* 582–590.

8 Zucker, K. J., Bradley, S. J., Owen-Anderson, A., Singh, D., Blanchard, R., & Bain, J. (2011). Puberty-blocking hormonal therapy for adolescents with gender identity disorder: A descriptive clinical study. *Journal of Gay & Lesbian Mental Health, 15,* 58–82.

9 Ibid., note 7.

10 T. D., Steensma, personal communication, February 3, 2020.

11 Referrals to GIDS, 2014–15 to 2018–19. (2019). The Tavistock and Portman NHS Foundation Trust. Retrieved from https://gids.nhs.uk/number-referrals.

12 Doward, J. (2018, November 3). Gender identity clinic accused of fast-tracking young adults. *Guardian.* Retrieved from https://www.theguardian.com/society/2018/nov/03/tavistock-centre-gender-identity-clinic-accused-fast-tracking-young-adults.

13 Churcher Clarke, A., & Spiliadis, A. (2019). "Taking the lid off the box": The value of extended clinical assessment for adolescents presenting with gender identity difficulties. *Clinical Child Psychology and Psychiatry, 24,* 338–352.

14 The role of the GP in caring for gender-questioning and transgender patients: RCGP position statement. (2019). Retrieved from: https://www.rcgp.org.uk/media/Files/Policy/A-Z-policy/2019/RCGP-transgender-care-position-statement-june-2019.ashx?la=en.

15 de Vries, A. L. C., Steensma, T. D., Doreleijers, T. A. H., & Cohen-Kettenis, P. T. (2011). Puberty suppression in adolescents with gender identity disorder: A prospective follow-up study. *Journal of Sexual Medicine, 8,* 2276–2283.

16 Ibid., Chapter 4, note 18.

17 Hembree, W. C., Cohen-Kettenis, P. T., Gooren, L., Hannema, S. E., Meyer,

W. J., Murad, M. H., . . . T'Sjoen, G. G. (2017). Endocrine treatment of gender-dysphoric/gender-incongruent persons: An Endocrine Society clinical practice guideline. *Journal of Clinical Endocrinology & Metabolism, 102,* 1–35.

18 Hough, D., Bellingham, M., Haraldsen, I. R. H., McLaughlin, M., Rennie, M., Robinson, J. E., . . . Evans, N. P. (2017). Spatial memory is impaired by peripubertal GnRH agonist treatment and testosterone replacement in sheep. *Psychoneuroendocrinology, 75,* 173–182.

19 Cohen, D., & Barnes, H. (2019, July 22). Transgender treatment: Puberty blockers study under investigation. BBC News. Retrieved from https://www.bbc.com/news/health-49036145.

20 Michael Laidlaw, an endocrinologist in California, discovered this through a Freedom of Information Act request regarding changes that one research team, operating on a five-year, $5.7 million National Institutes of Health research grant, made to their study protocol in 2017. Retrieved from https://docs.wixstatic.com/ugd/3f4f51_a929d049f7fb46c7a72c4c86ba43869a.pdf.

21 Haas, A. P., Rodgers, P. L., & Herman, J. L. (2014). *Suicide attempts among transgender and gender non-conforming adults: Findings of the National Transgender Discrimination Survey.* Williams Institute.

22 Doward, J. (2019, February 23). Governor of Tavistock Foundation quits over damning report into gender identity clinic. *Guardian.* Retrieved from https://www.theguardian.com/society/2019/feb/23/child-transgender-service-governor-quits-chaos.

23 Ibid., note 1.

24 Aitken, M., Steensma, T. D., Blanchard, R., VanderLaan, D. P., Wood, H., Fuentes, A., . . . Zucker, K. (2015). *Journal of Sexual Medicine, 12,* 756–763.

25 Health considerations for LGBTQ youth. (2019). Centers for Disease Control and Prevention. Retrieved from https://www.cdc.gov/healthyyouth/disparities/smy.htm.

26 Updated: Brown statements on gender dysphoria study. (2019, March 19). Retrieved from https://www.brown.edu/news/2019-03-19/gender.

27 For the full interview, see Kay, J., Delmar, D., & Soh, D.W. (2018, December 20). Gender dysphoria 101 with Susan Bradley. *Wrongspeak.*

28 Zucker, K. J., Bradley, S. J., Owen-Anderson, A., Singh, D., Blanchard, R. & Bain, J. (2011). Puberty-blocking hormonal therapy for adolescents with gender identity disorder: A descriptive clinical study. *Journal of Gay & Lesbian Mental Health, 15,* 58–82.

29 Robertson, J. D. (2019, March 22). Honoring a queen—The Marsha Johnson story. *Velvet Chronicle.* Retrieved from https://thevelvetchronicle.com/honoring-a-queen-the-marsha-johnson-story/.

30 Robertson, J. D. (2017, June 4). Remembering Stormé—The woman of color who incited the stonewall revolution. *Huffington Post.* Retrieved from https://

www.huffpost.com/entry/remembering-storm%C3%A9-the-woman-who-incited-the-stonewall_b_5933c061e4b062a6ac0ad09e.

31 Pornhub 2019 year in review. (2019). Retrieved from https://www.pornhub.com/insights/2019-year-in-review.

32 Dhejne, C., Öberg, K., Arver, S., & Landén, M. (2014). An analysis of all applications for sex reassignment surgery in Sweden, 1960–2010: Prevalence, incidence, and regrets. *Archives of Sexual Behavior, 43,* 1535–1545.

33 Lockwood, S. (2019, October 5). "Hundreds" of young trans people seeking help to return to original sex. *Sky News.* Retrieved from https://news.sky.com/story/hundreds-of-young-trans-people-seeking-help-to-return-to-original-sex-11827740.

Myth #6: No Differences Exist Between Trans Women and Women Who Were Born Women

1 We're not renaming the vagina. (2018, August 21). Retrieved from https://www.healthline.com/health/lgbtqia-safe-sex-guide/response#1.

2 American Cancer Society. (2020). Key statistics for prostate cancer. Retrieved from https://www.cancer.org/cancer/prostate-cancer/about/key-statistics.html.

3 Miksad, R. A., Bubley, G., & Church, P., Sanda, M., Rofsky, N., Kaplan, I., & Cooper, A. (2006). Prostate cancer in a transgender woman 41 years after initiation of feminization. *Journal of the American Medical Association, 296,* 2316–2317.

4 Blair, K. L., & Hoskin, R. A. (2018). Transgender exclusion from the world of dating: Patterns of acceptance and rejection of hypothetical trans dating partners as a function of sexual and gender identity. *Journal of Social and Personal Relationships, 7,* 2074–2095.

5 Gilligan, A. (2018, September 2). Unisex changing rooms put women in danger. *Sunday Times.* Retrieved from https://www.thetimes.co.uk/edition/news/unisex-changing-rooms-put-women-in-danger-8lwbp8kgk.

6 For examples, see Wood, M. (2016, June 3). 6 men who disguised themselves as women to access bathrooms. *Daily Signal.* Retrieved from https://www.dailysignal.com/2016/06/03/6-examples-highlight-serious-problems-with-obamas-bathroom-rule/.

7 Hellen, N. (2019, October 20). Police forces let rapists record their gender as female. *Sunday Times.* Retrieved from https://www.thetimes.co.uk/article/police-forces-let-rapists-record-their-gender-as-female-d7qtb7953.

8 First UK transgender prison to open. (2019, March 3). BBC News. Retrieved from https://www.bbc.com/news/uk-47434730.

9 Brean, J. (2018, August 2). Forced to share a room with transgender woman in Toronto shelter, sex abuse victim files human rights complaint. *National Post.* Retrieved from https://nationalpost.com/news/canada/kristi-hanna-human-rights-complaint-transgender-woman-toronto-shelter.

10 Ingle, S. (2019, September 24). IOC delays new transgender guidelines after scientists fail to agree. *Guardian.* Retrieved from https://www.theguardian.com/sport/2019/sep/24/ioc-delays-new-transgender-guidelines-2020-olympics.

Myth #7: Women Should Behave Like Men in Sex and Dating

1 Trivers, R. L. (1972). Parental investment and sexual selection. In B. Campbell (Ed.), *Sexual selection and the descent of man: 1871–1971* (pp. 136–179). Chicago: Aldine.

2 Buss, D. M. (1994). The strategies of human mating. *American Scientist, 82,* 238–249.

3 Greiling, H., & Buss, D. M. (2000). Women's sexual strategies: The hidden dimension of extra-pair mating. *Personality and Individual Differences, 28,* 929–963.

4 For additional information, see Blake, K. R., Bastian, B., Denson, T. F., Grosjean, P., & Brooks, R. C. (2018). Income inequality not gender inequality positively covaries with female sexualization on social media. *Proceedings of the National Academy of Sciences of the United States of America, 115,* 8722–8727.

5 For example, see Singh, D., Dixson, B. J., Jessop, T. S., Morgan, B. & Dixson, A. F. (2010). Cross-cultural consensus for waist-hip ratio and women's attractiveness. *Evolution and Human Behavior, 31,* 176–181.

6 Aharon, I., Etcoff, N., Ariely, D., Chabris, C. F., O'Connor, E., & Breiter, H. C. (2001). Beautiful faces have variable reward value: fMRI and behavioral evidence. *Neuron, 32,* 537–551.

7 Buss, D. M., & Dedden, L. A. (1990). Derogation of competitors. *Journal of Social and Personal Relationships, 7,* 392–422.

8 Aveline, M. (2016, September 26). Top modelling agent says male models "suffer big pay gap" compared to women. BBC News. Retrieved from http://www.bbc.co.uk/newsbeat/article/37456449/top-modelling-agent-says-male-models-suffer-big-pay-gap-compared-to-women.

9 Gul, P., & Kupfer, T. R. (2018). Benevolent sexism and mate preferences: Why do women prefer benevolent men despite recognizing that they can be undermining? *Personality and Social Psychology Bulletin, 45,* 1–16.

10 Campbell, A. (2008). The morning after the night before: Affective reactions to one-night stands among mated and unmated women and men. *Human Nature, 19,* 157–173.

11 Roberts, G. M. P., Newell, F., Simões-Franklin, & Garavan, H. (2008). Menstrual cycle phase modulates cognitive control over male but not female stimuli. *Brain Research, 1224,* 79–87.

Myth #8: Gender-Neutral Parenting Works

1 Colorado children's hospital drops gender from ID wristbands to make patients

"feel comfortable." (2018, September 28). *Global News.* Retrieved from https://globalnews.ca/news/4496612/childrens-hospital-gender-id-wristbands/.

2 Planned Parenthood Ottawa. [PPOttawa]. (2017, May 17). PPO supports the right of every person to determine their identity for themselves. We fully support implementation of #BillC16. [Tweet]. Retrieved from https://twitter.com/PPOttawa/status/864921424611483649.

3 Cherney, I. D., Kelly-Vance, L., Gill Glover, K., Ruane, A., & Oliver Ryalls, B. (2003). The effects of stereotyped toys and gender on play assessment in children aged 18–47 months. *Educational Psychology: An International Journal of Experimental Educational Psychology, 23,* 95–106.

4 Maglaty, J. (2011, April 7). When did girls start wearing pink? *Smithsonian Magazine.* Retrieved from https://www.smithsonianmag.com/arts-culture/when-did-girls-start-wearing-pink-1370097/.

5 Pasterski, V. L., Geffner, M. E., Brain, C., Hindmarsh, P. Brook, C., & Hines, M. (2005). Prenatal hormones and postnatal socialization by parents as determinants of male-typical toy play in girls with congenital adrenal hyperplasia. *Child Development, 76,* 264–278.

6 Lai, M.-C., personal communication, December 13, 2017.

7 Connellan, J., Baron-Cohen, S., Wheelwright, S., Batki, A., & Ahluwalia, J. (2000). Sex differences in human neonatal social perception. *Infant Behavior and Development, 23,* 113–118.

8 Todd, B. K., Barry, J. A., & Thommessen, S. A. O. (2016). Preferences for "gender-typed" toys in boys and girls aged 9 to 32 months. *Infant and Child Development, 26,* 1–14.

9 Hines, M. (2010). Sex-related variation in human behavior and the brain. *Trends in Cognitive Sciences, 14,* 448–456.

10 Hassett, J. M., Siebert, E. R., & Wallen, K. (2008). Sex differences in rhesus monkey toy preferences parallel those of children. *Hormones and Behavior, 54,* 359–364.

11 For more details, see Colapinto, J. (2004, June 3). What were the real reasons behind David Reimer's suicide? *Slate.* Retrieved from https://slate.com/technology/2004/06/why-did-david-reimer-commit-suicide.html.

12 Reiner, W. G., & Gearhart, J. P. (2004). Discordant sexual identity in some genetic males with cloacal exstrophy assigned to female sex at birth. *New England Journal of Medicine, 350,* 333–341.

Myth #9: Sexology and Social Justice Make Good Bedfellows

1 Mulvey, L., (1975). Visual pleasure and narrative cinema. *Screen, 16,* 6–18.

2 Buss, D. M., & von Hippel, W. (2018). Psychological barriers to evolutionary psychology: Ideological bias and coalitional adaptations. *Archives of Scientific Psychology, 6,* 148–158.

3 For the full interview, see Kay, J., Delmar, D., & Soh, D. W. (2018, November 9). Academic mobbing with Jonathan Haidt, Bret Weinstein, and Rebecca Tuvel. *Wrongspeak.*

4 Hawkins, S., Yudin, D., Juan-Torres, J., & Dixon, D. (2018). *Hidden tribes: A study of America's polarized landscape* (New York: More in Common).

5 Academics are being harassed over their research into transgender issues. (2018, October 16). *Guardian.* Retrieved from https://www.theguardian.com/society/2018/oct/16/academics-are-being-harassed-over-their-research-into-transgender-issues.

6 Peyser, A. (2020, January 27). "Far-left agitprop for pre-K tots: What NYC schools have come to." *New York Post.* Retrieved from https://nypost.com/2020/01/27/far-left-agitprop-for-pre-k-tots-what-nyc-schools-have-come-to/.

7 Dickson, C. (2018, March 7). "Got privilege?" B.C. school district under fire after launching anti-racism campaign. CBC News. Retrieved from: https://www.cbc.ca/news/canada/british-columbia/racism-campaign-school-district-74-1.4566779.

8 Humphreys, A. (2017, June 28). Meet "Sam," a transgender toy that's teaching kids about gender fluidity. *Mashable.* Retrieved from https://mashable.com/video/transgender-educational-toy-gender/.

9 Trans puppet named Julian to help young kids exploring gender issues (2018, September 6). *Canadian Press.* Retrieved from https://www.cbc.ca/news/canada/montreal/trans-puppet-julian-helps-kids-explore-gender-questions-1.4812357.

10 Hsu, K. J., Rosenthal, A. M., Miller, D. I., & Bailey, J. M. (2016). Who are gynandromorphophilic men? Characterizing men with sexual interest in transgender women. *Psychological Medicine,* 46, 819–827.

11 Bailey, J. M. (2019). How to ruin sex research. *Archives of Sexual Behavior, 48,* 1007–1011.

12 Blair, K. (2016). Ethical research with sexual and gender minorities. In A. E. Goldberg (Ed.), *The SAGE Encyclopedia of LGBTQ Studies* (375–380). Thousand Oaks, CA: SAGE.

13 Hill, T. P. (2018, September 7). Academic activists send a published paper down the memory hole. *Quillette.* Retrieved from https://quillette.com/2018/09/07/academic-activists-send-a-published-paper-down-the-memory-hole/.

14 Hurst, G. (2017, September 23). Bath Spa university bars research into transgender surgery regrets. *Times.* Retrieved from https://www.thetimes.co.uk/article/bath-spa-university-bars-research-into-transgender-surgery-regrets-ddxxlbfzh.

15 Special report: Online activists are silencing us, scientists say. (2019, March 18). Reuters. Retrieved from http://news.trust.org/item/20190313104914-k085q/.

16 Jebson Moore, K. (2019, February 17). A witch-hunt on Instagram. *Quillette.* Retrieved from https://quillette.com/2019/02/17/a-witch-hunt-on-instagram/.

INDEX